5th Symposium on the Mechanics of Slender Structures (MoSS2015)

Journal of Physics: Conference Series
Volume 721

Northampton, United Kingdom
21 – 22 September 2015

Editors:

Stefan Kaczmarczyk
Jose Manoel Balthazar
Ignacio Herrera Navarro

ISBN: 978-1-5108-2457-7
ISSN: 1742-6588

Printed from e-media with permission by:

Curran Associates, Inc.
57 Morehouse Lane
Red Hook, NY 12571

Some format issues inherent in the e-media version may also appear in this print version.

Printed by Curran Associates, Inc. (2016)

For permission requests, please contact the Institute of Physics at the address below.

Institute of Physics
Dirac House, Temple Back
Bristol BS1 6BE UK

Phone: 44 1 17 929 7481
Fax: 44 1 17 920 0979

techtracking@iop.org

Additional copies of this publication are available from:

Curran Associates, Inc.
57 Morehouse Lane
Red Hook, NY 12571 USA
Phone: 845-758-0400
Fax: 845-758-2633
Email: curran@proceedings.com
Web: www.proceedings.com

Table of contents

Volume 721

5th Symposium on the Mechanics of Slender Structures (MoSS2015)
21–22 September 2015, Northampton, UK

Accepted papers received: 23 May 2016
Published online: 2 June 2016

Preface

Papers

5th Symposium on Mechanics of Slender Structures 2015 (MoSS2015)

S Kaczmarczyk[1], J M Balthazar[2], I Herrera Navarro[3]

[1]The University of Northampton, UK
[2]Aeronautics Technological Institute, Brazil
[3]University of Extremadura, Spain

E-mail: stefan.kaczmarczyk@northampton.ac.uk

This issue of Journal of Physics: Conference Series contains sixteen selected technical papers that were presented at the 5th Symposium on the Mechanics of Slender Structures (MoSS2015) and show a range of different approaches used in the analysis of slender structures. The meeting was held in Northampton, UK, from 21st to 22nd September 2015. This conference runs under the auspices of the Institute of Physics Applied Mechanics Group and forms a continuation of a successful meeting series on the Mechanics of Slender Structures first held in Northampton, UK, in 2006, followed by the events hosted in Baltimore, USA in 2008, in San Sebastian, Spain in 2010, and in Harbin, China in 2013.

The remit of the Symposium on the Mechanics of Slender Structures series involves a broad range of scientific areas. Applications of slender structures include terrestrial, marine and space systems. Moving elastic elements such as ropes, cables, belts and tethers are pivotal components of many engineering systems. Their lengths often vary when the system is in operation. The applications include vertical transportation installations and, more recently, space tether propulsion systems. Traction drive elevator (lift) installations employ ropes and belts of variable length as a means of suspension, and also for the compensation of tensile forces over the traction sheave. In cranes and mine hoists, cables and ropes are subject to length variation in order to carry payloads. Tethers experiencing extension and retraction are important components of offshore and marine installations, as well as being proposed for a variety of different space vehicle propulsion systems based on different applications of momentum exchange and electrodynamic interactions with planetary magnetic fields. Furthermore, cables and slender rods are used extensively in civil engineering; in cable-supported bridges, guyed masts and long-span roofs of buildings and stadia. Suspended cables are also applied as electricity transmission lines. Chains are used in various power transmission systems that include such mechanical systems as chain drives and chain saws. Moving conveyor belts are essential components in various material handling installations. Other structures such as pipelines, plates, beams, mechanical linkages, and DNA structures also fall into this category.

Tall buildings and towers represent another important class of slender structures. In the modern high-rise built environment tall buildings have increased height and slenderness as well as reduced weight. Such structures are designed to withstand a broad range of external loads such as strong wind and seismic excitation. However, they are prone to structural vibrations and complex resonance phenomena that cause damage, affect their occupants and modular installations such as vertical transportation/lift systems. The performance of these installations plays a significant role in the building operation and a holistic approach is needed in the analysis and design of the structural system.

This meeting brought together experts from various fields: structural mechanics, thermo-mechanics, dynamics, electrodynamics, vibration and control, structural health monitoring, artificial intelligence, materials science and applied mathematics to discuss the current state of research as well as rising trends and direction for future research in the area of mechanics of slender structures. The event is aimed at improving the understanding of structural and thermo-mechanical properties and behaviour of slender structures. The papers presented at the conference covered analytical, numerical and experimental research in various applications of slender structures. The conference programme was arranged around the following five keynote lectures:

Keynote Lecture 1: Chaotic and noisy-chaotic dynamics of slender structures by Jan Awrejcewicz, Lodz University of Technology (Poland).

Keynote Lecture 2: Influences on lifetime of wire ropes in traction lifts by Wolfram Vogel, University of Chemnitz (Germany).

Keynote Lecture 3: Complex Dynamics of Drill-String and BHA by Marian Wiercigroch, University of Aberdeen (UK).

Keynote Lecture 4: A tour of nonlinear dynamic analysis of elastic structures excited by non-ideal sources (NIS) by Jose Manoel Balthazar, Aeronautics Technological Institute (Brazil).

Keynote Lecture 5: Dynamic response of mechanical systems to impulse process stochastic excitations: Markov approach by Radoslaw Iwankiewicz, Hamburg University of Technology (Germany)

Keynote Lecture 6: Modal Analysis of Sailplane and Transport Aircraft Wings Using the Dynamic Stiffness Method by Ranjan Banerjee, City University London (UK)

The articles presented in this volume are arranged alphabetically by the first author details and are extended peer-reviewed versions of the papers presented at the conference. The Editors and Organizing Committee gratefully acknowledges support received from the co-sponsoring institutions and would like to thank the authors for their hard work and high quality contributions.

Editors:

Stefan Kaczmarczyk, The University of Northampton, UK (Editor)
José Manoel Balthazar, Aeronautics Technological Institute, Brazil (Co-Editor)
Ignacio Herrera Navarro, University of Extremadura, Spain (Co-Editor)

The International Organizing Committee:

Jose Manoel Balthazar, Aeronautics Technological Institute, Brazil
Ignacio Herrera Navarro, University of Extremadura, Spain (Co-Chair)
Stefan Kaczmarczyk, University of Northampton, UK (Co-Chair)
Seyed Mirhadizadeh, University of Northampton, UK
Professor Weidong Zhu, University of Maryland BC, USA.

THE UNIVERSITY OF NORTHAMPTON

IOP Institute of Physics Stress and Vibration Group

CIBSE LIFTS GROUP

BINDT THE BRITISH INSTITUTE OF NON-DESTRUCTIVE TESTING

Endorsed by: The ASME Technical Committee on Vibration and Sound (TCVS)

21-22 September 2015, Highgate House Conference Centre, Creaton, Northampton.

Journal of Physics: Conference Series **721** (2016) 011002 doi:10.1088/1742-6596/721/1/011002

Peer review statement

All papers published in this volume of *Journal of Physics: Conference Series* have been peer reviewed through processes administered by the proceedings Editors. Reviews were conducted by expert referees to the professional and scientific standards expected of a proceedings journal published by IOP Publishing.

Mechanical discrete simulator of the electro-mechanical lift with n:1 roping

F J Alonso[1a], **I Herrera**[2a]

[a]Escuela de Ingenierías Industriales, Universidad de Extremadura, Avda. de Elvas s/n, 06006 Badajoz, Spain

[1]fjas@unex.es, [2]iherrera@unex.es

Abstract. The design process of new products in lift engineering is a difficult task due to, mainly, the complexity and slenderness of the lift system, demanding a predictive tool for the lift mechanics. A mechanical ad-hoc discrete simulator, as an alternative to 'general purpose' mechanical simulators is proposed. Firstly, the synthesis and experimentation process that has led to establish a suitable model capable of simulating accurately the response of the electromechanical lift is discussed. Then, the equations of motion are derived. The model comprises a discrete system of 5 vertically displaceable masses (car, counterweight, car frame, passengers/loads and lift drive), an inertial mass of the assembly tension pulley-rotor shaft which can rotate about the machine axis and 6 mechanical connectors with 1:1 suspension layout. The model is extended to any n:1 roping lift by setting 6 equivalent mechanical components (suspension systems for car and counterweight, lift drive silent blocks, tension pulley-lift drive stator and passengers/load equivalent spring-damper) by inductive inference from 1:1 and generalized 2:1 roping system. The application to simulate real elevator systems is proposed by numeric time integration of the governing equations using the Kutta-Meden algorithm and implemented in a computer program for ad-hoc elevator simulation called ElevaCAD.

1. Introduction

The analysis of the dynamic behavior of the elevator car system plays an important role in elevator engineering and superior ride quality of elevators is demanded nowadays. In particular, vibration in the low frequency range must be investigated in the modern design of elevator systems. International standard encourage the development of uniform, reliable and precise measurement and processing techniques to be applied within the elevator industry [1]. It is a challenging task to develop a precise and reliable tool/software in order to simulate the dynamic characteristics of the elevator system. In addition to research, education and training applications could be improved with a tool like that.

The use of general purpose software packages [2-4] is the unique alternative when elevator engineers need to simulate the dynamics of the elevator, at present, if an elevator test tower or any other facility for experimentation is not available. The implementation of a mechanical model in a software package entails the modeling of the system components. The basic assumptions on the elevator system dynamics are postulated in order to get a precise response in standard elevator trip and residential building elevators in a first stage. So the previous work [5] was devoted to study the elevator dynamics by testing in real installations and take conclusions on the trends of the system response and critical components parameters. It was detected that fine variations of those parameters greatly influenced the overall response, so a list of such parameters and specific test procedures were developed and done [6].

The main and critical components of the elevator system are the passengers. As it is well known, the behavior of passengers is very complicated factor to accommodate in the elevator model [7-9]. Further on, the passengers are the most important "mechanical components" of the elevator system. The task for modeling the passengers could be even more challenging than that of the elevator system itself. Recently, elevator models and measurement procedures have been researched [10-11] to determine the damping and stiffness characteristics of passengers.

Dedicated specialist multibody dynamics software packages like Adams [2], SolidWorks [3] and Working Model [4] can be applied to study the dynamic response of the elevator. Implicitly, the use of simulators means idealization so some mechanic model of the elevator must be defined. This paper deals with the process and justification to derive an optimized valid mechanical model of the elevator system as a whole, based on rope dynamics analysis and testing results on in-situ elevators.

2. The elevator system

In a typical elevator system the car frame is suspended by sling and the car is mounted within the car frame on isolation blocks (see figure 1). The elevator is inside a building hole known as shaft and supported by a number of solutions to the building structure. In figure 1 the lift drive, the diverter pulley and its bed is isolated from the building wrought by silent blocks. Minimum torque requirement makes quite common to suspend also a counterweight by the other end of the sling.

The car frame and the counterweight are forced by the guide rails to move substantially in the vertical direction, without twist, by roller guides or sliding guide shoes components. In the elevator system, vibrations in both lateral and vertical directions are of interest. Lateral and vertical displacements of the car are important for comfort and noise control but vertical displacements are essential. Vertical vibrations are determinant to the dynamic response, safety, durability, fatigue, maintenance and comfort of the elevator system.

The car, car-frame, and counterweight are generally made of metallic pieces and fixed to each counter-part by bolts and nuts or welding techniques. Special attention is devoted to tighten the pieces to avoid noise during travel. Every of these components almost constitute an ideal rigid solid. The six degrees of freedom of a rigid solid are reduced to only one, thus assuming ideal sliding junctions between car-frame or counterweight and the guide rails.

The components of the elevator system has been studied separately and derivation of a simplified one degree of freedom model was supported by experiments.

The response of the suspension ropes is essential to understand and model the elevator system. The sling is generally made by a number n_R of tension equalized metallic ropes disposed in series, fixed to the car-frame by one end and to the counterweight by the other end. The contact between the rope and traction sheave and diverter pulleys is specially critical, not only for the neccesity to produce the desired movement of the car but its control, safety, and adherence response too.

3. The proposed model

According to figure 2 we assign the subscript (*CW*) to the counterweight, (*FR*) to the car frame, (*CA*) to the car, (*PA*) to the weights/passengers travelled, and (*MA*) to the drive including its foundation. They all are considered rigid bodies. Each of these bodies is assigned one vertical degree of freedom y_{CW}, y_{FR}, y_{CA}, y_{PA} and y_{MA} (See figure 2).

Let's assign the variable y_{FR} for the vertical displacement of the car-frame and the variable y_{CW} for that of the counterweight. The car is fixed to the car-frame by four silent-blocks. This makes the car to be movable in the vertical direction. The variable y_{CA} is assigned. We also must consider the most important part of the elevator that the passengers or goods are. We assign for simplicity variable y_{PA} to the vertical displacement of the elevator continent.

The structure of the machine drive is composed of two rigid parts the drive stator-drive foundation and the shaft which integrates the traction sheave, rotor, brakes (*BR*) and the flywheel. The shaft can rotate almost ideally. So we conceive the shaft rotation α as one degree of freedom of the elevator system. The machine drive-foundation substructure is isolated from its base to avoid noise transmission to the building structure then is assigned an extra degree of freedom to its vertical displacement y_{MA}. The product of the shaft rotation α by the radius of the traction sheave R is the displacement of a peripherial point of the traction sheave αR which is used instead of α to get dimensional homogeneity of the displacement vector *[y]*. The drive brakes (*BR*) when actuated are contemplated in the model by an equivalent spring damper (*BR*) of stiffness k_{BR} and damping c_{BR}.

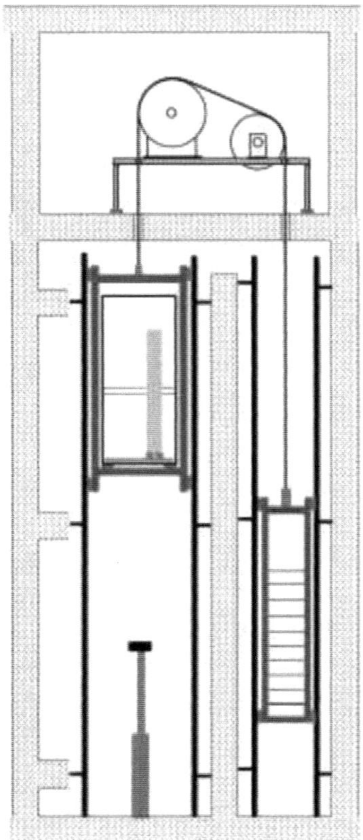

Figure 1. Elevator system scheme.

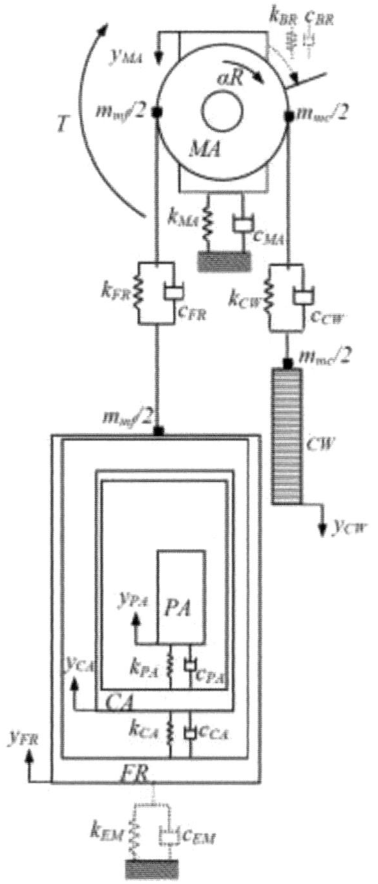

Figure 2. 6 d.o.f mechanical model of the elevator car system.

Simulation of the impact of the car-frame against the buffers or safety gear actuation is suggested by an equivalent spring/damper (*EM*) of stiffness k_{EM} and damping c_{EM}.

If the car is joined by several silent-blocks (or other type of similar low stiffness dampers) to the car frame, an equivalent system where the car and the car frame are connected by an equivalent spring/damper of stiffness k_{CA} and damping c_{CA} is suggested. If we consider that the car of mass m_{CA} is attached to the car frame by a number of isolation blocks located at each corner of the car floor (parallel configuration of the individual silent-blocks) the resulting stiffness and damping coefficients for the car are:

$$k_{CA} = n_{ib} k_{ib} ; c_{CA} = n_{ib} c_{ib} \tag{1}$$

where n_{ib} is the number of isolation blocks used in the car and k_{ib} and c_{ib} are the stiffness and viscous damping coefficients of a single isolation block.

It is conventionally assumed undamped condition in many analysis with ropes. The available decay data of damping tests on ropes [12] and specially designed tests previously researched [10-11] have revealed that rope undamping is not acceptable for the elevator's mechanical modelling we are driving. The suspension ropes possess some rigidity to bend when subjected to loads of the order of 1/12 times the one for reaching its elastic limit (12 is a typical safety factor for in-service three ropes suspensions for attaining fatigue criteria that EN-81-20 normalizes). That means that the composition of the rope influences its dynamics performance by means of a non constant damping factor ς which is affected to some extend by the suspended load.

The rope-pieces connecting the traction pulley and the car frame of mass m_{FR} is equivalent to a linear spring/damper of stiffness k_{FR} and damping c_{FR}, length l_f and a number of paralell suspension ropes n_R:

$$k_{FR} = n_R \frac{EA}{l_f(t)}; c_{FR} = 2\varsigma \sqrt{\frac{n_R EA}{l_f(t)} \cdot (m_{FR} + m_{CA} + m_{PA})} \quad (2)$$

where E is the apparent elastic modulus of a single suspension rope, A is its true sectional area, and ς is the damping factor of a single rope which it is assumed to be tensioned by a suspended mass of $(m_{FR} + m_{CA} + m_{PA})/n_R$. The damping factor can be computed from a series of tests with a single rope of the same type of any length and suspended mass and assumed constant.

Similarly, the rope-piece connecting the traction pulley and the counterweight of mass m_{CW} is equivalent to a linear spring/damper of constants:

$$k_{CW} = n_R \frac{EA}{l_c(t)}; c_{CW} = 2\varsigma \sqrt{\frac{n_R EA}{l_c(t)} \cdot m_{CW}} \quad (3)$$

where l_c is the length of the sling rope-piece, and ς is the damping factor of a single rope when subjected to tension by a suspended mass of m_{CW}/n_R.

The rope mass is also studied. The rope-piece mass connecting the frame to the traction sheave is denoted by m_{mf}, and the rope-piece mass connecting the counterweight to the traction sheave is denoted by m_{mc}. These two masses can be easily determined as a function of the rope density ρ, effective cross section A, the number of suspension ropes n_R and its instantaneous lengths $l_f(t)$ and $l_c(t)$, as follows:

$$m_{mf} = n_R \rho A l_f(t); m_{mc} = n_R \rho A l_c(t) \quad (4)$$

The simplified model joins half the mass of the sling frame to the car-frame mass. Then, the remaining half is joint to the periphery of the traction drive. The same mass distribution applies to the counterweight sling. In this way, the computed static deformations of the elevator components in the reference configuration of the system are in agreement with Strength of Materials formulas [14] without increasing model complexity [13].

Also, the drive-foundation silent blocks defines mainly the equivalent joint between the machine stator and the building structure/ground k_{MA} and c_{MA} assuming that if several massless springs joints are dispossed in series, the equivalent spring can be aproximated by the lowest individual stiffness.

The passengers are joined to the car mass, m_{CA}, through two stiffless mass components: the shoes (sh) and the covering car floor (fl). Both can be treated separately as ideal spring/dumper joints. Then, it is assumed [7,8] that the human body spring/damper in the vertical direction, the shoes and the floor spring/dampers are connected in series, and can be reduced to one ideal spring/damper joint whose constants are k_{1p} and c_{1p}, respectively.

$$\frac{1}{k_{1p}} = \frac{1}{k_{pa}} + \frac{1}{k_{sh}} + \frac{1}{k_{fl}}; \frac{1}{c_{1p}} = \frac{1}{c_{pa}} + \frac{1}{c_{sh}} + \frac{1}{c_{fl}} \quad (5)$$

where k_{1p} denotes the overall equivalent stiffness coefficient of one passenger in the vertical direction, k_{pa} the equivalent stiffness of one passenger in bear feet position, k_{sh} the shoes-stockings vertical stiffness coefficient of one passenger, and k_{fl} the vertical stiffness coefficient of the car floor. Similarly, c_{1p}, c_{pa}, c_{sh} and c_{fl} denote the overall equivalent vertical viscous damping for one passenger, the equivalent vertical viscous damping coefficients for the passengers, the shoes-stockings coefficients and the car floor coefficient.

Assuming that all the passengers travelling in the car move in phase relative to each other and have the same mass with the average passenger's mass being m_{1p}, the resulting coefficients k_{PA} and c_{PA} for all passengers in the car are:

$$k_{PA} = n_p k_{1p}; c_{PA} = n_p c_{1p}; m_{PA} = n_p m_{1p} \tag{6}$$

where n_P is the number of passengers travelling in the car. Similar rules apply if goods alone or mixed passengers/goods are carried by the car.

4. n:1 roping equations

Following 2nd Newton's law the governing equation of the system of rigid solids shown in figure 2 can be written and generalized to n:1 roping:

$$[M][\ddot{y}]+[C][\dot{y}]+[K][y]=[F] \tag{7}$$

where $[M]$ is the mass matrix, $[C]$ is the damping matrix, $[K]$ is the stiffness matrix, $[F]$ is the time dependent vector of applied forces and $[y]$ is the time displacement vector. The mass matrix has the following form:

$$[M]=\begin{bmatrix} \frac{I_{RM}}{R^2}+\frac{m_{mf}}{3}+\frac{m_{mc}}{3}+\sum_{d=1}^{n_d}\frac{I_d}{r_d^2} & \frac{m_{mf}}{6} & 0 & 0 & \frac{m_{mc}}{6} & 0 \\ \frac{m_{mf}}{6} & m_{FR}+\frac{m_{mf}}{3} & 0 & 0 & 0 & 0 \\ 0 & 0 & m_{CA} & 0 & 0 & 0 \\ 0 & 0 & 0 & m_{PA} & 0 & 0 \\ \frac{m_{mc}}{6} & 0 & 0 & 0 & m_{CW}+\frac{m_{mc}}{3} & 0 \\ 0 & 0 & 0 & 0 & 0 & m_{MA}+\frac{m_{mf}}{2}+\frac{m_{mc}}{2} \end{bmatrix};[y]=\begin{bmatrix}\alpha R \\ y_{FR} \\ y_{CA} \\ y_{PA} \\ y_{CW} \\ y_{MA}\end{bmatrix} \tag{8}$$

where I_{RM} is the shaft inertia which integrates the traction sheave, brakes, rotor and flywheel, R is the traction sheave radius, n_d is the number of diverter pulleys, I_d the inertia of the d-th diverter pulley, r_d is the radius of the d-th diverter pulley, m_{MA} is the machine mass including the traction sheave mass, m_{FR} is the car-frame mass, m_{CA} is the car mass, and m_{PA} is the mass of the passengers or goods.

The distributed mass that the slings rope-pieces are, has been included in the model by assembling the governing equations of the consistent mass approach [13]. This apply two both the rope-piece connecting the frame to the drive which mass is denoted by m_{mf}, and the rope-piece connecting the counterweight to the drive which mass is denoted by m_{mc}. These two masses were previously derived as a function of its instantaneous lengths $l_f(t)$ and $l_c(t)$ as shown in equation (4). The instantaneous lengths can be computed as a function of the roping ratio n, the instantaneous car frame position $y_{CA}(t)$, the maximum trip distance from 0-floor to n-th floor h, and the minimum free installation lengths of the frame rope-piece s_{0f} (car stopped in the n-th floor) and the counterweight rope-piece s_{0c} (for the car stopped in the 0 floor) by:

$$l_f(t) = s_{0f} + n(h - y_{CA}(t))$$
$$l_c(t) = s_{0c} + n y_{CA}(t) \tag{9}$$

The minimum free installation lengths s_{0f} and s_{0c} can be estimated from h_f and h_w shown in the installation layout (see figure 3) by $s_{0f} \approx nh_f$ and $s_{0c} \approx nh_w$. It was considered that the small pieces of rope of the slings in contact with the traction sheave and diverter pullies may not be included in the distributed mass of the slings m_{mf} and m_{mc}, nor influence its stiffness k_{fr} and k_{cw} and damping c_{fr} and c_{cw}. See how the linear rope speed in the counterweight sling $\dot{l}_c(t)$ is positive and how, in the frame sling $\dot{l}_f(t)$, is negative in an ascending trip $(\dot{y}_{CA}(t) \geq 0)$:

$$n\dot{y}_{CA}(t) = -\dot{l}_f(t) = \dot{l}_c(t) \tag{10}$$

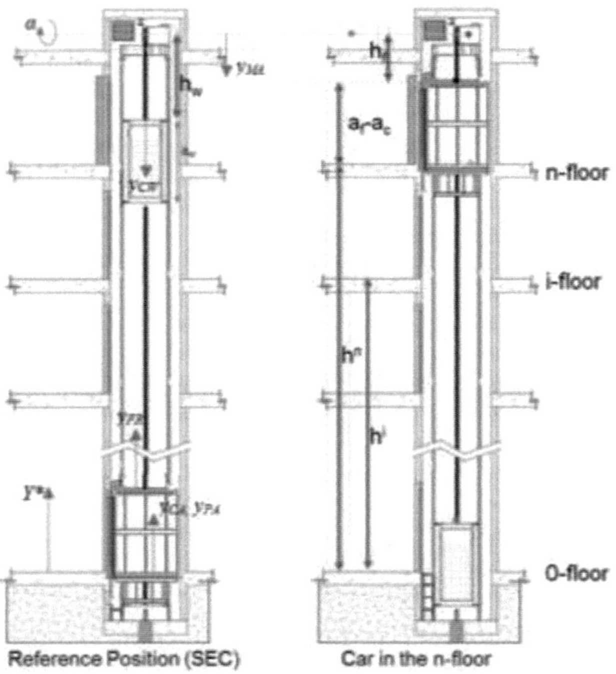

Figure 3. Installation parameters and origin of displacement of the system proposed.

and how its absolute value is n times greater according to the $n:1$ roping of the elevator if conventional differential no resistance to rotation pullies are used.

The stiffness matrix $[K]$ of the elevator system has the form:

$$[K] = \begin{bmatrix} k_{CW}+k_{FR}+k_{BR} & -k_{FR} & 0 & 0 & -k_{CW} & k_{CW}-k_{FR} \\ -k_{FR} & k_{FR}+k_{CA}+k_{EM} & -k_{CA} & 0 & 0 & k_{FR} \\ 0 & -k_{CA} & k_{CA}+k_{PA} & -k_{PA} & 0 & 0 \\ 0 & 0 & -k_{PA} & k_{PA} & 0 & 0 \\ -k_{CW} & 0 & 0 & 0 & k_{CW} & -k_{CW} \\ k_{CW}-k_{FR} & k_{FR} & 0 & 0 & -k_{CW} & k_{MA}+k_{FR}+k_{CW} \end{bmatrix} \tag{11}$$

where:

$$\frac{1}{k_{FR}}=\frac{1}{n^2}\left(\frac{1}{n_R\dfrac{EA}{l_f(t)}}+\sum_{i=1}^{n_{df}}\frac{1}{k_{di}}+\frac{1}{k_{fe}}\right);\frac{1}{c_{FR}}=\frac{1}{n^2}\left(\frac{1}{2\varsigma\sqrt{\dfrac{n_R EA}{l_f(t)}\cdot(m_{FR}+m_{CA}+m_{PA})}}+\frac{1}{c_{fe}}\right) \tag{12a}$$

$$\frac{1}{k_{CW}}=\frac{1}{n^2}\left(\frac{1}{n_R\dfrac{EA}{l_f(t)}}+\sum_{i=1}^{n_{dc}}\frac{1}{k_{di}}+\frac{1}{k_{ce}}\right);\frac{1}{c_{CW}}=\frac{1}{n^2}\left(\frac{1}{2\varsigma\sqrt{\dfrac{n_R EA}{l_c(t)}\cdot m_{CW}}}+\frac{1}{c_{ce}}\right) \tag{12b}$$

are generalized from previous expressions (2-3) due to the facts that in a $n:1$ roping installation the rope extension of the frame sling Δl_f, for example, is equal to n times the frame displacement y_{FR}, the suspended car weight P is n times the force exerted by the rope sling N ($P=nN$) and that the equivalent stiffness is the car weight divided by its displacement ($k_{FR}=P/y_{FR}$). Then the "in series" effect of the end terminals and the diverter pullies, present in most elevator installations, are added. The terms k_{fe} and c_{fe} are the stiffness and damping of the frame sling end terminals, respectively; k_{ce} and c_{ce} are the stiffness and damping of the counterweight sling end terminals, respectively, n_{df} and n_{dc} are the number of diverter pulleys in the frame sling and counterweight sling, respectively; and k_{di} is the stiffness of the i-th pulley which deviates the rope sling.

The effect in the stiffness of the i-th diverter pulley is shown in figure 4. The rope is wrapped around the pulley with the angle of wrap β_i. The isolated pulley is tensioned by the rope traction P/n. If the pulley is fixed at the middle of a hinged-hinged axle beam of length l_{di}, 2nd moment of area I_{di} and modulus of elasticity E_{di}, the deflection of the beam/axle at the mid span f is given by:

$$f=2\cdot\frac{P}{n_R}\sin\left(\frac{\beta_i}{2}\right)\frac{l_{di}^3}{48E_{di}I_{di}}\Rightarrow\frac{1}{k_{di}}=\frac{\Delta l}{P/n}=\frac{2f\sin(\beta_i/2)}{P/n}=\frac{4\sin^2\left(\dfrac{\beta_i}{2}\right)l_{di}^3}{48E_{di}I_{di}} \tag{13}$$

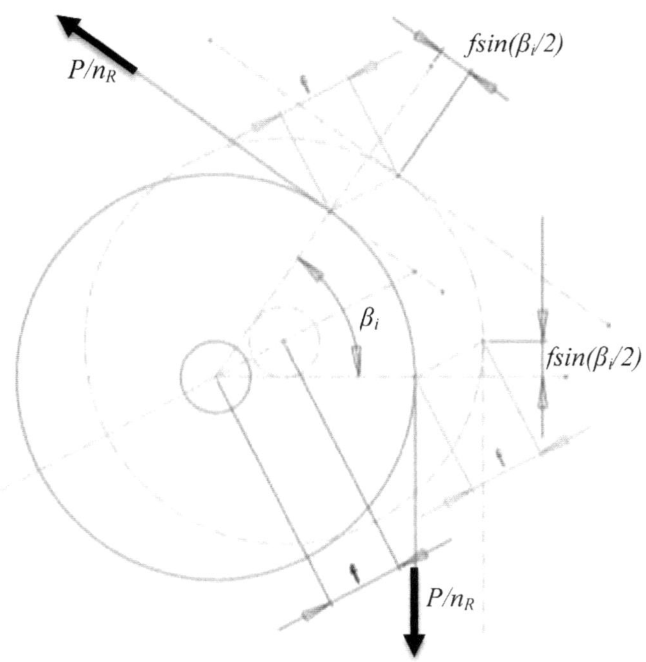

Figure 4. Derivation of the wrap angle β_i diverter pulley stiffness from the deflection f.

During normal operation both spring/damper for emergency actuation k_{EM}, c_{EM} are kept to zero (no actuation of the buffer nor the safety gear nor car-frame impact against a guide due to guide missalignment occur).

Finally, the applied forces vector $[F]$ under the presence of gravity of intensity g has the form:

$$[F]=\begin{bmatrix} \dfrac{T}{R}+\dfrac{(m_{mc}-m_{mf})g}{2} \\ -(m_{FR}+m_{mf}/2)g \\ -m_{CA}g \\ -m_{PA}g \\ (m_{CW}+m_{mc}/2)g \\ (m_{MA}+m_{mf}/2+m_{mc}/2)g \end{bmatrix} \tag{14}$$

where T is the applied torque by the drive on the traction sheave which includes the motor torque and the torque applied by the brakes that makes it the rotor to stop rotation. The force exerted by the frame sling mass m_{mf} and counterweight sling mass m_{mc} on the frame and the counterweight, respectively, include the factor ½ for equal static components deformation of the real system and the modeled discrete system in the reference configuration which is coincident with that of the system just after the putting on tension process. Then, the remaining mass of the slings $(m_{mf}+m_{mc})/2$ fixed to the machine (MA) has to contribute to the vertical displacement and gravity torque of the machine. The total static torque exerted on the traction sheave is the sum of the momentum applied by the drive torque T and the gravity torque: $(m_{mc}-m_{mf})gR/2$.

5. Results and Conclusions

Now we are ready to compare the performance of the various roping systems at the functional differences in the elevator installation. This is based on the analysis of various parameters in expressions (7-14) and has not involved solving the equations (7) which is undertaken in [14].

Table 1. Comparison of mechanical parameters for a number of roping systems. N_1: permissible tension of one rope. T_0: minimum drive torque required. k_{1fe}: spring constant of 1 rope end. k_{d1}: spring constant of 1 diverter pulley.

Roping	1:1	2:1	4:1	n:1
$(m_{FR}+m_{CA}+m_{PA}+m_{mf})g/N_1$	n_R	$2n_R$	$4n_R$	$n\cdot n_R$
$(m_{FR}+m_{CA}+m_{PA}+m_{mf}-m_{CW}-m_{mc})gR/T_0$	1	2	4	n
$\alpha R/y_{FR}$ (permanent regime)	1	2	4	n
(Necessary rope length)/h	n_R	$2n_R$	$4n_R$	$n\cdot n_R$
$[k_{FR}$(rope end factor)$]/k_{1fe}$	n_R	2^2n_R	4^2n_R	$n^2\cdot n_R$
$[k_{FR}$(rope factor top drive)$]/(EA/h)$	n_R	$2n_R$	$4n_R$	$n\cdot n_R$
$[k_{FR}$(rope factor bottom drive)$]/(EA/h)$	$n_R/2$	$2^2n_R/3$	$16n_R/5$	$n^2n_R/(n+1)$
Min number of div pulleys $(\beta_i=\pi)$	0	2	6	$2(n-1)$
$[k_{FR}$ $(\beta_i=\pi$ div pulleys factor$)]/k_{d1}$	-	4	16/3	n^2/n_{df}
Min number of div pulleys $(\beta_i=\pi/2)$	0	4	12	$4(n-1)$
$[k_{FR}$ $(\beta_i=\pi/2$ div pulleys factor$)]/k_{d1}$	-	1	16/6	n^2/n_{df}
Min $[M](1,1)$(div pulleys factor)$/(I_d/r_d^2)$	-	2	6	$2(n-1)$

From table 1 above it is obvious that increasing roping ratio n reduces the minimum torque of the machine drive and the permissible load of the required rope but the necessary rope length is increased to the same amount. The permissible load of the required rope can also be reduced by increasing the number of ropes n_R. The roping ratio n influences, mainly, the dynamics of the installation. The angular speed of the motor drive is increased by n and the startup and shutdown of the elevator will relax by the inherent increase in the number of diverter pulleys. However, the passengers comfort which is influenced by the stiffness of the suspension ropes remains almost constant for equal permissible tension of the rope design.

The formulas for generalized roping of the elevator has been derived. The proposed 6 d.o.f. system has been able to accommodate the performance of complex *n:1* roping elevator systems. The simplicity of the proposed model is adequate to sensitivity analysis and assist the design process in lift engineering and education applications. The proposed model is implemented in ad-hoc software called ElevaCAD that implements the Kutta-Medem algorithm to get the solution of the governing equations by numeric time iteration to any excitation of the system and speed control strategy [14].

References

[1] ISO/TC 178 2012 Lifts Measurement of lift ride quality *ISO 18738:2012*(en) (2012)

[2] MSC Software Adams *Multibody Dynamics Simulation* Newport Beach. California. USA.

[3] Dassault Systemes SolidWorks Simulation *Motion Analysis* Cedex. France.

[4] Design Simulation Technologies Working Model 2D *http://www.design-simulation.com/wm2d/* Canton. USA.

[5] Herrera I 2008 Structure of VIBRELEV-1G: Simulation Software for the Elevator Dynamic Response *Elevator Technology 17* 160-168 IAEE ISBN 978.965.90338.5.0

[6] Herrera I 2010 Mechanical Modelling and Simulation Lift Systems: the Influence of the Dynamic Behaviour of Slender Structural Components *Proceedings of MoSS 2010 Symposium on the mechanics of slender structures* ISBN.978.84.608.1100.8

[7] Herrera I and Kaczmarczyk S 2009 The Assessment of Vibration Absorption Capacity of Elevator's Passengers *7th International conference on Modern Practice in stress and Vibration Analysis* Journal of Physics Conference Series 181.

[8] Herrera I, Su H and Kaczmarczyk S 2013 Influence of the number of passengers on the dynamic response of an elevator car *11th International Conference on Vibration Problems. Z.* Dimitrovová et al (eds) Lisbon Portugal 9-12 septembrer 2013.

[9] Herrera I, Su H and Kaczmarczyk S 2015 Influence of the Load Occupancy Ratio on the Dynamic Response of an Elevator Car System *Applied Mechanics and Materials* Vol 706 (2015) pp128-136 doi:10.4028/www.scientific.net/AMM.706.128.

[10] Herrera I, Su H and Kaczmarczyk S 2010 Investigation into the damping and stiffness characteristics of an elevator car system *Applied Mechanics and Materials* Trans Tech Publications 24-25 (2010) 77-82.

[11] Herrera I, Su H and Kaczmarczyk S 2014 Modelling and estimation of damping and stiffness characteristics of lift passengers *AIP Conference proceedings* 1637 395(2014) doi:10.1063/1.4904604.

[12] Wehking K.H. et al. Cited in pp 96-99 *Wire Ropes Tension, Endurance, Reliability* by Klaus Feyrer. Springer.

[13] Herrera I and Kaczmarczyk S 2015 Simple models for rope substructures mechanics: Application to electro-mechanical lifts *Presented to MoSS 2015*

[14] Herrera I and Romero E 2015 Software design to calculate and simulate the mechanical response of electromechanical lifts *Presented to MoSS 2015*

Journal of Physics: Conference Series **721** (2016) 012002 doi:10.1088/1742-6596/721/1/012002

Modelling of thermal behaviour of iron oxide layers on boiler tubes

J D Angelo[1], A Bennecer[2], S Kaczmarczyk[2], P Picton[2]

[1] Universidade Federal do ABC, São Paulo, Brasil.
[2] University of Northampton, Northampton, U.K.

Abstract. Slender boiler tubes are subject to localised swelling when they are expose to excessive heat. The latter is due to the formation of an oxide layer, which acts as an insulation barrier. This excessive heat can lead to microstructural changes in the material that would reduce the mechanical strength and would eventually lead to critical and catastrophic failure. Detecting such creep damage remains a formidable challenge for boiler operators. It involves a costly process of shutting down the plant, performing electromagnetic and ultrasonic non-destructive inspection, repairing or replacing damaged tubes and finally restarting the plant to resume its service.
This research explores through a model developed using a finite element computer simulation platform the thermal behaviour of slender tubes under constant temperature exceeding 723 °K. Our simulation results demonstrate that hematite layers up to 15μm thickness inside the tubes do not act as insulation. They clearly show the process of long term overheating on the outside of boiler tubes which in turn leads to initiation of flaws.

1. Introduction

Boilers and other heavy industrial equipment operating at high temperatures and pressure are design for a finite life [1]. The lifetime of boilers and steam generators are between 20 to 30 years [2]. In recent years, heavy industrial structures have aged beyond their design lifetime. Indeed, most boiler tubes have been operating for more than 30 years. Therefore, frequent inspections are required to ensure that the boilers are still safe to use and it is cost effective to keep them running for longer periods. **Figure 1** shows the areas of one aqua tube wall subject to overheating.

The temperature inside the water wall tubes can reach 943 °K. However, super heaters and reheaters can operate at temperatures of nearly 1163 °K. At these temperatures, microstructural changes occur in ferritic steel structures but they do not pose a substantial risk of failure [2]. One study conducted by Graham and Waleed [3] at Saudi Aramco on 101 cases of boilers tubes concluded that 24% of tubes such as furnaces fail due long term overheating and 6% of tubes such as super heaters fail due to short term overheating. The distribution of damage mechanisms is depict in **Figure 2**.

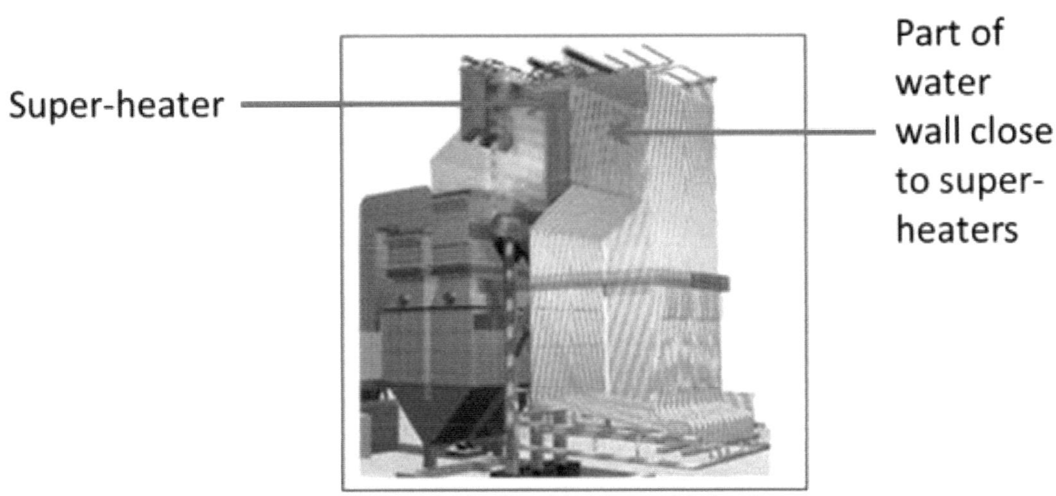

Figure 1: Hot areas shown in red are subject to overheating in a boiler [4].

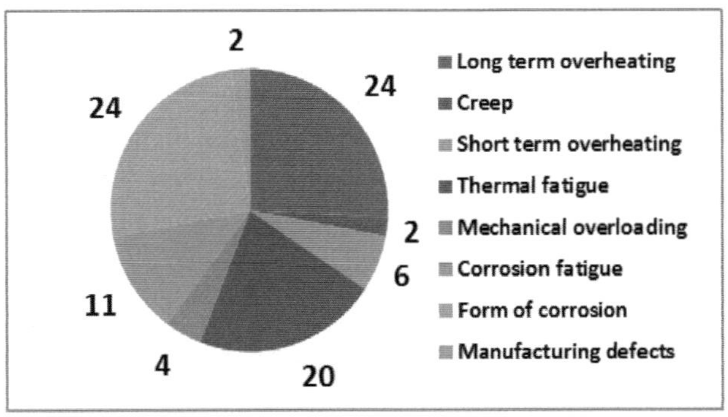

Figure 2: Boiler tubes common damage mechanisms.

Overheating is associated with the formation of different oxide layers with different thicknesses over the internal surface of the tubes where less than 3% Cr is present. At temperatures below 833 °K and high partial pressure of O_2 this film is made of two layers: 1) a magnetite (Fe_2O_3) layer and 2) a hematite (Fe_2O_4) layer. At temperatures higher than 873 °C an additional layer of iron monoxide (FeO) also appears.

The effects of the oxidation layers on the mechanical behaviour of materials manifest themselves in a reduction of the tube cross section, which increases the localised stress. The film growth causes an insulation of this section leading to a temperature increase within the cross section and acceleration of material degradation. Also, spalling of oxide scales can be beneficial to reduce the aforementioned effects, but can conversely result in tube overheating if it becomes entrapped in the system, thus reducing the flow rate in this section [5].

The presence of two or more layers of dissimilar materials, which present a mismatch between the coefficients of thermal expansion, induces stress. As consequence, a delamination or more likely a crack would occur. In this paper, we study the effects of long term overheating due to the formation of oxide layers inside boiler tubes. We use finite element method in order to model the thermal behaviour of the tube under long term overheating. The study is conduct under stationary conditions to reflect real boiler conditions.

2. Methodology

The data of operating pressure, design pressure, design temperature and feed water temperature used in our study is sourced from a VU-60 boiler [6] (see Table 1). The structure is made of seamless carbon steel and is 6000 mm long with approximately 60 mm outside diameter and 50 mm inside diameter. It is subject to heating in excess of 450 °C in one longitudinal half. In reality, the tube length, diameter and thickness depend on the work pressure and the quantity of steam height generated.

Table 1: Boiler operational parameters [6]

Model	VU-60
Operating pressure	6.423 [MPa]
Design pressure	6.816 [MPa]
Design temperature	728.15 [°K]
Feed water temperature	463.15 [°K]

One-half of the tube is expose to ambient temperature while the other half is expose to a combustion chamber. After a certain height, steam starts to evaporate from boiled water inside the tube. It is in this part of the tube where the oxide layer is likely to form and grow causing long term overheating due to the reduction in heat transfer between tube wall and steam. As the diffusion rate of these oxide layers raises exponentially with the temperature new alloys are develop slowing the growth rate. The heat transfer through a steel tube wall with iron oxide layer can be considered unidirectional as shown in **Figure 3**.

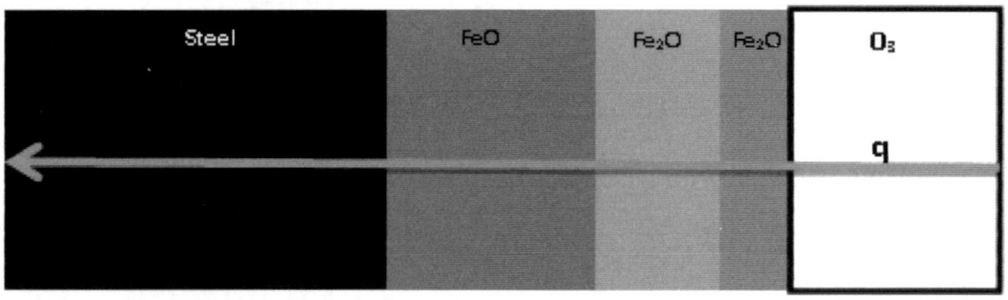

Figure 3: Heat transfer through iron oxide layer [7].

The oxide layers growing are due to the partial pressure of oxygen over the tube wall. If not all oxygen formed over the tube wall is consumed the oxygen bubble tries scape from the oxide layer. The movement of bubbles trying scape from the oxide layer creates porosity inside the oxide layer. Each oxide layer has a different energy of formation. This difference of energy of formation causes some discontinuity from one layer to another layer. In this model, we assume a unidirectional heat flux, there is a perfect adherence between the oxide layers and tube wall, and there is no porosity inside the oxides [7]. With this assumption, the heat quantity that flows through the oxide layers and tube wall is given by the equation:

$$\frac{q}{A} = k\frac{dT}{dx} \tag{1}$$

Where:
q=heat transfer rate,
k=material thermal conductivity,
A=material cross section,

T=temperature in degrees Kelvin, and,
x=material thickness.

In a stationary study, equation (1) reduces to:

$$\frac{q}{A} = k \quad (2)$$

Assuming also that the oxide layer is composed of 90% wustite, 8% magnetite and 2% hematite [3], we impose a maximum oxide layer thickness of 1000 μm. Therefore, the wustite layer will have a thickness of 900 μm, the magnetite layer will be 80 μm and the hematite layer will be 20 μm.

We also consider the design temperature (728 °K) as the upper limit and the inlet water temperature (463 °K) as the steam temperature limit. We start our study by considering a new tube without any oxide layer and then we gradually add hematite layers of 10 μm, 15 μm and 20 μm to explore the effects on heat distribution on the tube wall. The material properties used in this study are show in **Table 2**.

Table 2: Thermo-mechanical properties.

Material	k (W/mK)	ρ (kg/m^3)	C_p (J/kgK)
Carbon steel	44.5	7850	475
FeO	3.2	7750	725
Fe_3O_4	1.5	5600	800
Fe_2O_3	1.2	4900	980

COMSOL Multiphysics®[1] ver. 4.4 is used with heat transfer in solids for modelling these phenomena.

3. Results

The results obtained by isothermal contour analysis shows that the oxide layers growing act as insulation, making the tube wall facing the combustion chamber hotter than the tube wall inside the refractory wall. We also observe that as the oxide layers grow the internal surface of the tubes facing the combustion chamber becomes colder than the outside surface in the combustion chamber.

The isothermal contour of carbon steel tubes without oxide layers show that the temperature varies from 813 °K on the tube surface in contact with combustion chamber to 693 °K on the steam in contact with tube.

When a hematite layer grows over the internal surface of tube, the temperature profile varies from 783oK to 503 °K for a 10 μm hematite oxide layer. For a 15 μm hematite oxide layer the temperature varies from 783 °K to 635 °K and for a 20 μm hematite oxide layer the temperature varies from 783 °K to 582 °K.

When a magnetite oxide layer is added to a hematite oxide layer the temperature profile varies from 783 °K to 529 °K for 60 μm and 80 μm of magnetite. This indicates, as observed in the field, that at a certain magnetite thickness this oxide acts as a protective layer.

When we added a wustite oxide layer to the magnetite oxide layer the temperature profile varies from 783 °K to 476 °K for 450 μm and 675 μm wustite and from 783 °K to 529 °K for a 900 μm wustite.

Our results, using isothermal contours from COMSOL Multiphysics®, show that there is a critical thickness of oxide layer where the temperature profile changes. These changes in the temperature profile could be associated with coarsening pores and crack growth inside the oxide layer, as shown in **Figure 4** [7].

---

[1] Comsol Multiphysics is a trademark from COMSOL Inc.

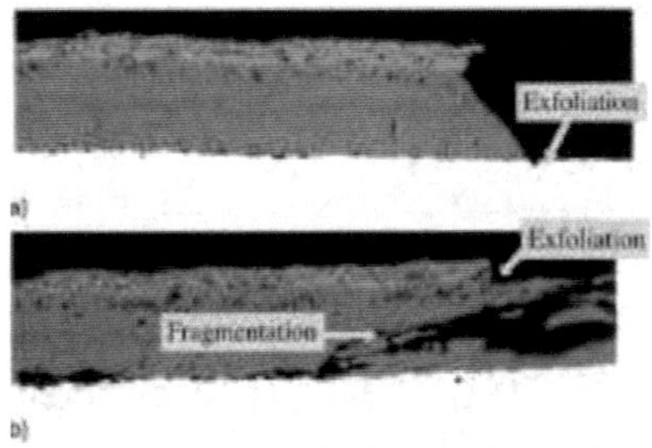

Figure 4: Picture showing coarse pores and exfoliation in hematite oxide layer [7].

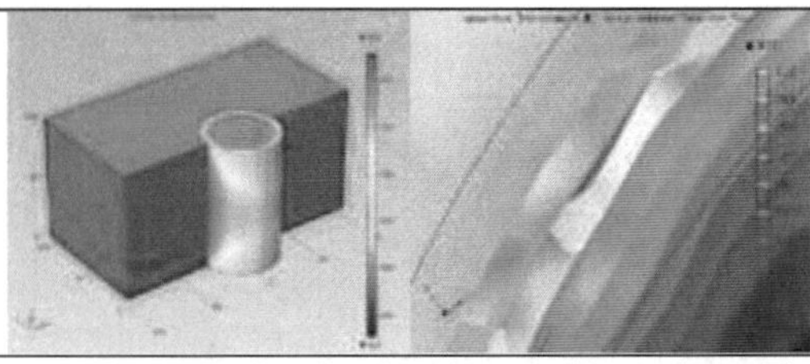

Figure 5: Heat distribution in a boiler tube without any oxide layers and a close-up of the heat distribution.

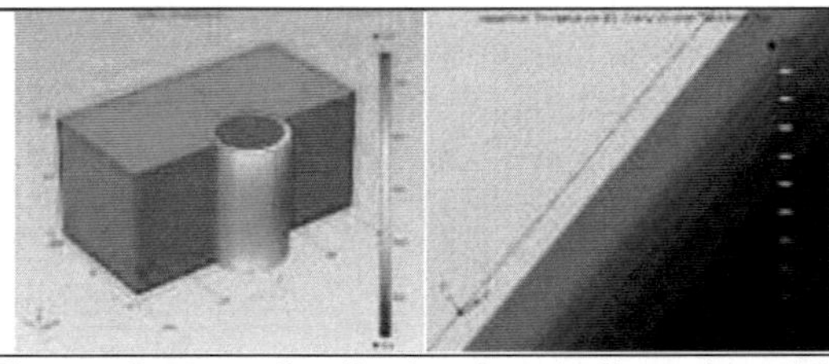

Figure 6: Heat distribution in a boiler tube with a 10 μm hematite layer and a close-up of the heat distribution in the hematite layer.

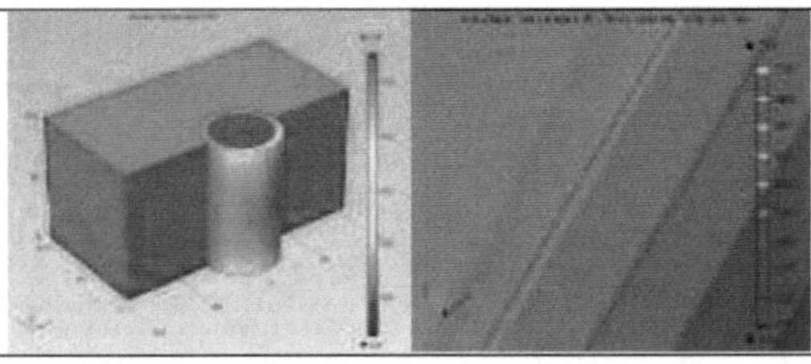

Figure 7: Heat distribution in a boiler tube with a 15 μm hematite layer and a close-up of the heat distribution in the hematite layer.

Figure 8: Heat distribution in a boiler tube with a 20 μm hematite layer and a close-up of the heat distribution in the hematite layer.

Figure 9: Heat distribution in a boiler tube with a 40 μm magnetite and 10 μm hematite layer and a close-up of the heat distribution in the layers.

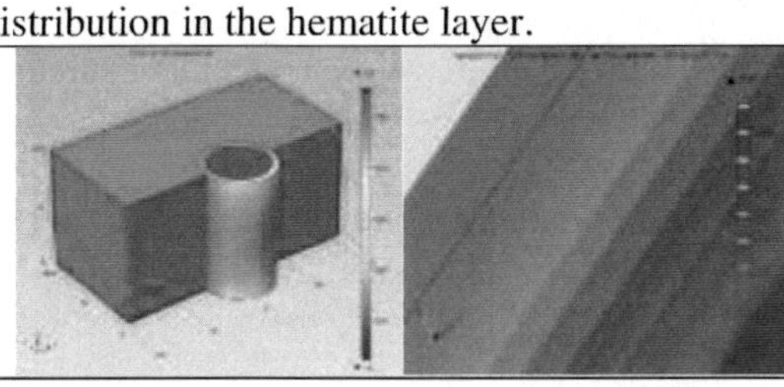

Figure 10: Heat distribution in a boiler tube with a 60 μm magnetite and a 15 μm of hematite layer and a close-up of the heat distribution in the layers.

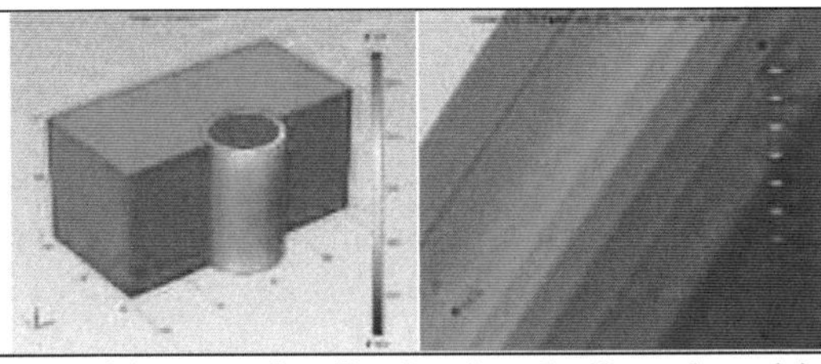

Figure 11: Heat distribution in a boiler tube with an 80 µm magnetite and a 20 µm hematite layer and a close-up of the heat distribution in the layers.

Figure 12: Heat distribution in a boiler tube with a 450 µm wustite, magnetite and a 10 µm hematite layer and a close-up of the heat distribution in the layers.

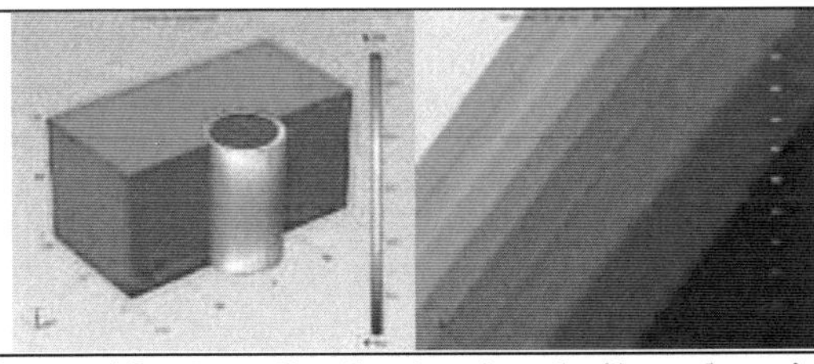

Figure 13: Heat distribution in a boiler tube with a 675 µm wustite, 60 µm magnetite and a 15 µm of hematite layer and a close-up of the heat distribution in the layers.

Figure 17: Heat distribution in a boiler tube with a 900 µm wustite, an 80 µm magnetite and a 20 µm hematite layer and a close-up of heat distribution in the layers.

Figures 17 to **20** show the refractory effects of each iron oxide layer as it grows

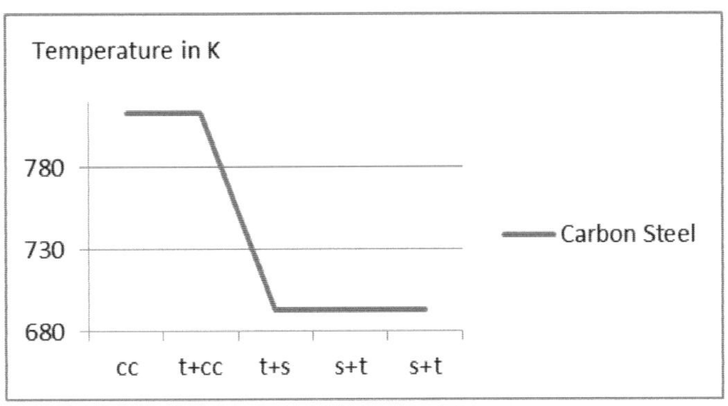

Figure 17: Graph showing a carbon steel without oxides.

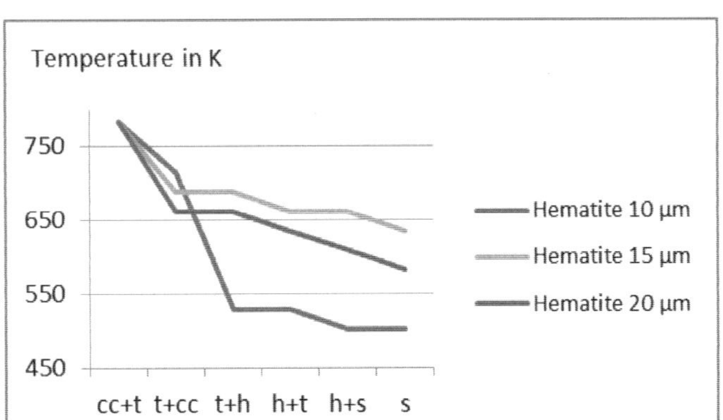

Figure 18: Graph showing the refractory effect of a hematite layer growing.

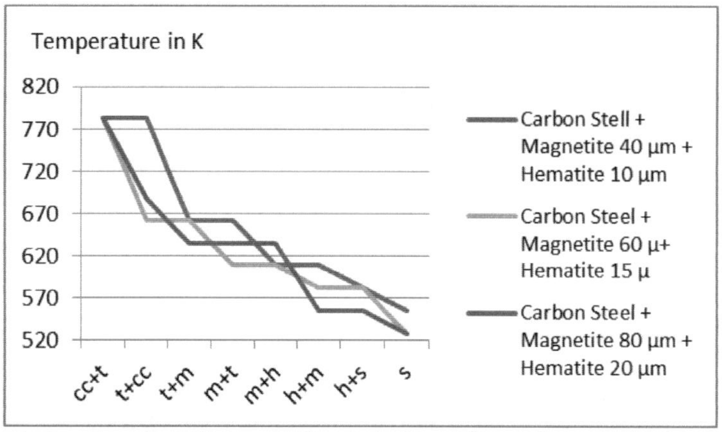

Figure 19: Graph showing the refractory effect of a hematite and magnetite layer growing.

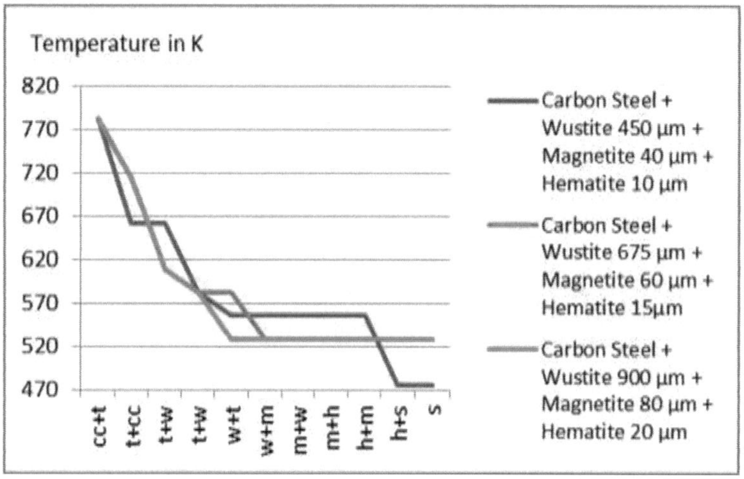

Figure 20: Graph showing the refractory effect of a wustite, magneite and hematite layer growing.

3. Conclusion.

The simulation of oxides growing indicates that:

1 –A thinner layer of hematite up to 10 µm is less refractory than a thicker hematite layer over 10 µm. It seems be due to thinner layers of hematite being more compact than thicker layers which permits oxygen to penetrate through hematite oxide to form magnetite.

2 – Magnetite and hematite layers are less refractory than hematite.,

3 – Having a wustite, magnetite and hematite layer is the worst composition of oxides layers and the most refractory. This composition affects the heat distribution causing microstructural changes that leads tube failure. The refractory behaviour of this composition of oxide layers could be associated with the fact that wustite is an iron and chromium oxide. This composition, due to the chromium content, is more refractory.

4- An effective control of oxygen content in the feed water, the operating temperature and the use of NDT technique's which is able to measure hematite thickness could prevent tubes failure.

Acknowledgments

J D Angelo would like acknowledge to CNPq – Brazil, for the research grant, which funded his research.

References

[1] Singer J G 1991 Combustion, Fossil Power. ABB Combustion Engineering, Windsor, Connecticut.

[2] French D N Microstructural Degradation. The National Board of Boiler and Pressure Vessel Inspector web site, Classic Articles, available at www.nationalboard.org/classics.

[3] Graham R, Lobley and Waleed l. Al-Otaibi, 2008, Diagnosing Tube Failures Related to Overheating, Advanced Materials Research, **41-42**, pp 175-181.

[4] Catalogue Energia, Sermatec-Zanini, available in http://media.sermatec.com.br/uploads/produtos_arquivo/2_catalogo_sermateczanini_energia_2.pdf.

[5] Bruno Reis Cardoso et al, 2012, J. of Mater. Res. Technol, **1(2),** 109-116.

[6] Martin Torres and Rafael Colas, 2000, A model for heat conduction through the oxide layer of steel during hot rolling, Journal of Materials Processing Technology, **105 (3),** 258-263.

[7] Program on Technology Innovation: Oxide Growth and Exfoliation on Alloys Exposed to Steam. EPRI, Palo Alto, CA:2007. 1013666.

Non-symmetric and chaotic vibrations of Euler-Bernoulli beams under harmonic and noisy excitations

J Awrejcewicz[1], N P Erofeev[2] and V A Krysko[3]

[1] Department of Automation, Biomechanics and Mechatronics, Lodz University of Technology, 1/15 Stefanowski St., 90-924 Lodz and Department of Vehicles, Warsaw University of Technology, 84 Narbutta Str., 02-854 Warsaw, POLAND
awrejcew@p.lodz.pl
[2,3] Department of Mathematics and Modeling, Saratov State Technical University, Politehnicheskaya 77, 410054 Saratov, RUSSIAN FEDERATION, tak@san.ru

Keywords: geometric non-linearity, Euler-Bernoulli beam, white noise, noise-induced transitions, non-symmetric vibrations.

Abstract. In this paper we study non-linear dynamics of flexible Euler-Bernoulli beams subjected to harmonic load and white noise. We report that in the case of continuous mechanical systems like the studied Euler-Bernoulli beams the action of white noise yieldes novel and unexpected phenomena, i.e. symmetric transversal loads and symmetric boundary conditions imply non-symmetric beam vibration. In addition, regions of non-symmetric beam vibrations for two values of white noise intensity are given and the charts of regular and chaotic vibrations of the studied beams are presented

1. Introduction

Noisy nature of a medium may essentially change the behavior of a dynamical system. This novel type of non-equilibrium transition can be induced by a noise [1]. On the other hand, in deterministic systems a few scenarios of transition from regular to chaotic dynamics are detected and reasonably well studied. Let us mention the Ruelle-Takens scenario [2] where only three frequencies are involved to produce a chaotic attractor. In this scenario noisy (chaotic) system behavior has been associated with an occurrence of a strange attractor arose after three successive period-doubling bifurcations. Many researchers studied simple non-linear physical systems which have been able to generate internal noise (chaos) [3]. Scenarios associated with period doubling (Andronov-Hopf) bifurcations as well as with the intermittency phenomena have been detected and extensively analyzed [4]. However, noisy nature of a medium is capable of inducing much richer and unexpected non-linear dynamical phenomena than the above mentioned. We show that the methods applied to study deterministic classical phase transition can be extended to study noisy systems. Thus, it becomes possible to apply the classical theory of chaos to analyze noise-induced transitions to chaotic vibrations by numerous numerical approaches. In the following paper we are aimed at applying the classical theory of non-linear multi-dimensional deterministic dynamical systems putting emphasis on the noise-induced regular and chaotic dynamics. There exist numerous papers and books devoted to study deterministic chaos exhibited by simple non-linear oscillators. These studies have been directly implemented to analyze also continuous systems by taking their single-mode approximation and by applying the

Bubnov-Galerkin approach. However, results of this research can be treated as the qualitative ones since the increase in the number of modes may induce their interaction and the arose vibrations may be significantly different from those offered by one degree-of-freedom (1-DoF) approximations.

The problem of non-linear PDEs governing dynamics of the Euler-Bernoulli beams taking into account the geometric non-linearity, which is the subject of our paper, cannot be solved in a closed analytical way. Therefore, numerical methods are applied including FEM (Finite Element Method), FDM (Finite Difference Method), TFE (Time Finite Element), and the Bubnov-Galerkin approaches based on higher order of approximations. In general, the numerical solution requires discretizations of both spatial and time domains. In result, the problem is reduced to a large set of ODEs (Ordinary Differential Equations) with two numerical control parameters, i.e. the spatial grid size and the time step size.

Bar-Yoseph et al. [5] pointed out that nonlinear spatio-temporal dynamical systems governed by non-linear PDEs, may display periodic and aperiodic solutions in time and space including solitons, shock waves, ultrasubharmonic, quasi-periodic and chaotic attractors.

They have employed the space-time spectral element method to solve a simply supported modified Euler-Bernoulli beam undergoing external forced lateral vibrations. It has been shown that the characteristics of the first mode yielded the first period doubling bifurcations, the onset of chaotic strange attractor, and then a stable periodic solution. It has been also reported that the form and size of the first mode strange attractor is sensitive to the time step size, the spatial grid size, and the polynomial order.

Bokhari et al. [6] have studied symmetries and integrability of a fourth-order Euler-Bernoulli equation. They have employed the Lie and Noether symmetry analysis. In our paper we have also addressed this problem, but we have shown that our symmetric system may exhibit a break of symmetry due to action of the white noise.

The literature review does not exhibit research devoted to an interplay of chaotic dynamics of the structural members with white Gaussian noise. However, forced random vibration of beam structures is interest in both theory and application and has been widely documented in the existing literature.

Dunne and Ghanbari [7] have compared predicted extreme exceedance possibilities associated with experimental measurements of highly non-linear damped-damped beam vibrations driven by band-limited white noise. It has been shown that by exploiting the Weissman estimator the accurate exceedance probabilities from relatively small amounts of measured data can be obtained.

The stochastic bifurcation exhibited by one flexible beam subject to axial Gauss white noise excitation has been analyzed in reference [8]. The stochastic stability of the trivial system solution has been estimated using the Lyapunov exponents and boundary classifications.

Lan and Qin [9] have employed the method of the energy harvesting from the horizontal coherent resonance of a vertical cantilever beam subjected to the vertical beam excitation. It has been shown that the horizontal coherence resonance can be detected by introducing a vertical white noise.

The existence of a compact random attractor for the floating beam equation with strong damping and white noise has been investigated by Ling Xu and Ma [10].

In the reference [11] global bifurcations and chaotic dynamics of a cantilever beam under axial harmonic load and transversal excitation at its free end have been reported. Rich chaotic dynamics exhibited by plates and shells taking into account their geometric non-linearity has been illustrated and discussed in references [12-14], among other.

In this paper we study vibrations of geometrically non-linear Euler-Bernoulli beams as systems with many DOFs under external excitation in the form of white noise. Here the term "white" refers to white light which has the whole range of frequencies in the visible spectrum. The generalized correlation function governing white noise has the form of: $B(t)=\sigma^2\delta(t)$, where σ^2 is a positive constant and $\delta(t)$ denotes the delta (δ) function. Our study may have an impact on non-linear dynamics of plates and shells under action of the sound pressure, which corresponds to the excitations models in the form of Gaussian white noise.

2. Problem formulation

We study a single-layer beam treated as a 2D region of the R^2 space with Cartesian coordinate system introduced in the following way: we fix a median line at *z*=0, whereas the axis *OX* is directed from the left to the right along the median line and the remaining axis *OZ* is directed downwards, perpendicularly to *OX*. In the introduced coordinate system the beam is considered as a two-dimensional area Ω defined as follows: $\Omega = \{x \in [0,a]; -h \le z \le h\}$, $0 \le t \le \infty$. Here and below we use the following notation: $2h$ denotes the height of the beam and a stands for its length.

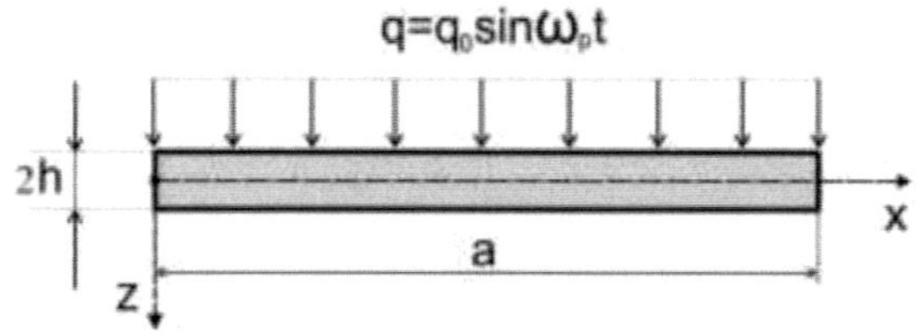

Figure 1. The considered Euler-Bernoulli beam

3. Mathematical model

A mathematical model of the beam has been developed on a basis of the Euler-Bernoulli hypothesis taking into account the non-linear relationship between the beam deformation and displacements in the von Kármán form [15]. The system of non-linear PDEs governing beam motion including the energy dissipation has the following form:

$$\begin{gathered} 2hE\left\{\frac{\partial^2 u}{\partial x^2} + L_3(w,w)\right\} - 2h\frac{\gamma}{g}\frac{\partial^2 u}{\partial t^2} = 0, \\ 2hE\left\{L_1(u,w) + L_2(w,w) - \frac{(2h)^2}{12}\frac{\partial^4 w}{\partial x^4}\right\} + q + \tilde{q} - 2h\frac{\gamma}{g}\frac{\partial^2 w}{\partial t^2} - 2h\varepsilon\frac{\gamma}{g}\frac{\partial w}{\partial t} = 0, \\ L_1(u,w) = \frac{\partial^2 u}{\partial x^2}\frac{\partial w}{\partial x} + \frac{\partial u}{\partial x}\frac{\partial^2 w}{\partial x^2}, \quad L_2(w,w) = \frac{3}{2}\frac{\partial^2 w}{\partial x^2}\left(\frac{\partial w}{\partial x}\right)^2, \quad L_3(w,w) = \frac{\partial^2 w}{\partial x^2}\frac{\partial w}{\partial x}, \end{gathered} \tag{1}$$

where $w(x,t)$ is the beam deflection, $u(x,t)$ stands for the beam displacement along the axis *OX*, ε is the damping coefficient, $q = q(x,t)$ denotes beam transversal load, $\tilde{q}(t)$ is the applied white noise in the form $\tilde{q} = w_n \cdot$(2.0 rand()/(RAND_MAX+1.0)-1.0), E is Young's modulus, ν denotes the Poisson coefficient, whereas ρ, γ stand for the density and the volumetric weight of the beam material, g – the gravity of Earth, w_n – white noise intensity. The following dimensionless quantities are introduced:

$$\bar{w} = \frac{w}{2h},\ \bar{u} = \frac{ua}{(2h)^2},\ \bar{x} = \frac{x}{a},\ \lambda = \frac{a}{2h},\ \bar{q} = q\frac{a^4}{(2h)^4 E},\ \bar{t} = \frac{t}{\tau},\ \tau = \frac{a}{c},\ c = \sqrt{\frac{Eg}{\gamma}},\ \bar{\varepsilon} = \varepsilon\frac{a}{c}. \tag{2}$$

Using (2) in (1) we get

$$\begin{gathered} \frac{\partial^2 u}{\partial x^2} + L_3(w,w) - \frac{\partial^2 u}{\partial t^2} = 0, \\ \frac{1}{\lambda^2}\left\{L_1(u,w) + L_2(w,w) - \frac{1}{12}\frac{\partial^4 w}{\partial x^4}\right\} - \frac{\partial^2 w}{\partial t^2} - \varepsilon\frac{\partial w}{\partial t} + q + \tilde{q} = 0, \end{gathered} \tag{3}$$

where bars above dimensionless quantities are omitted for simplification (we have fixed $\lambda = 50, \varepsilon = 1.0$).

In (3) we consider the following boundary conditions applied on the beam ends

$$w(0,t)=w(1,t)=u(0,t)=u(1,t)=\partial^2 w(0,t)/\partial x^2=\partial^2 w(1,t)/\partial x^2=0, \quad (4)$$

and the following initial conditions

$$w(x,t)\,|_{t=0}=\left.\frac{\partial w(x,t)}{\partial t}\right|_{t=0}=u(x,t)\,|_{t=0}=\left.\frac{\partial u(x,t)}{\partial t}\right|_{t=0}=0. \quad (5)$$

4. Methods of solution

In what follows we substitute the differential operators with respect to the variable x as well as their derivatives by the difference operators with approximation $O(h^2)$. Therefore, system (3) is reduced to the following set of ordinary differential equations (ODEs):

$$\ddot{w}_i+\varepsilon\dot{w}_i=q+\tilde{q}+\frac{1}{\lambda^2}\left\{L_{1,h}(u_i,w_i)+L_{2,h}(w_i,w_i)-\frac{1}{12}l_1(w_i)\right\},\quad \ddot{u}_i=L_{3,h}(w,w_i)+l_2(u_i),\quad i=0,\dots,n, \quad (6)$$

where n denotes a number of nodes regarding the space coordinate, $L_{1,h}$, $L_{2,h}$ and $L_{3,h}$ stand now for non-linear difference operators, $l_1(w), l_1(w_i)$ and $l_2(u_i)$ are linear difference operators given in a standard form. ODEs (6) together with boundary (4) and initial (5) conditions are derived through application of the FDM with approximation $O(h^2)$ and they are solved using the standard 4th order Runge-Kutta method. The carried out numerical experiments have shown that the FDM convergence regarding nodes number for $x\in[0;1]$ achieved for n=40 by the Runge-Kutta methods of 4th and 6th orders is the same. This is why further the 4th order Runge-Kutta method has been applied, since it requires less computational time.

5. Analysis of the simulation results

In order to study dynamics of non-linear beams subjected to harmonic load $q=q_0\sin(\omega_p t)$ and white noise we apply the so-called charts of beam vibration types versus the control parameters $\{q_0,\omega_p,w_n\}$. To get a point of this chart we have to analyze the frequency power spectrum as well as the Lyapunov exponents (LEs) for each set of $\{q_0,\omega_p,w_n\}$.

The developed algorithm allows to detect regions of periodicity, period doubling bifurcation regions, zones of quasi-periodicity as well as zones of chaos. Below we report time histories $w(t)$ in the middle of the beam for the beam length $x\in[0;1]$ divided into 40 parts. The time history refers to the time interval $t\in[0;1024]$ with the boundary conditions (4). To obtain the charts of vibration character depending on the control parameters $q_0\in[100;60000]$ with the division into 600 parts and $\omega_p\in[0.0003;12]$ with the division into 350 parts one needs to study $2.1\cdot10^5$ times the governing differential equations. In other words, to obtain the above mentioned chart each of the points from a set of $2.1\cdot10^5$ points has been analyzed with the help of the frequency power spectrum, the Morlet wavelet, modal phase portraits, autocorrelation functions, the Lyapunov exponents and the Poincarè maps $w_{t+T}(w_t)$, where T denotes a period of regular excitation.

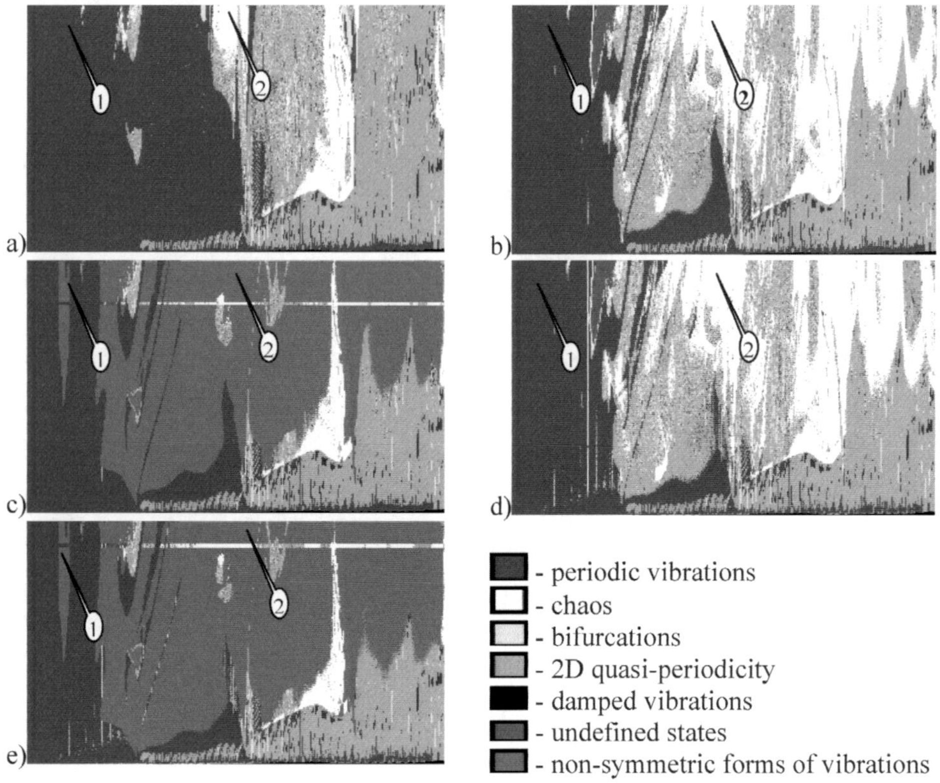

Figure 2. Charts of beam's vibration regimes for different noise excitation amplitude: a) w_n=0; b) w_n=0.01; c) w_n=0.1; d) w_n=1; e) w_n=10

In Fig. 2 the charts of vibration types in the case of white noise absence w_n=0 (Fig. 2a) are shown accompanied by the charts for two values of white noise amplitude w_n=0.01 and w_n=1 (Fig. 2b,d) as well as by zones of non-symmetric forms of vibrations for the white noise amplitude w_n=0.1 (Fig. 2c) and w_n=1 (Fig. 2e). We have chosen two points with the following coordinates: point 1 ($\omega_p = 1.0503, q_0 = 43300$), point 2 ($\omega_p = 5.5303, q_0 = 56700$). Tables 1, 2 report the wavelets, time histories, power spectra, Poincarè maps, phase portraits and snapshots of the beam deflections $w(x,t), x \in [0;1]$ at the given time instants for the point 1 and 2, respectively. The overall charts analysis shows that the presence of white noise significantly reduces zones of periodic vibrations but variations of the intensity of white noise have little effect on the change of zones of periodicity. It should be noted that an increase in the noise amplitude up to w_n=1 in the resonance zone yields an increase in periodic waves regions. It means that we may control periodic vibrations of the beam by adding white noise properly. It is particularly important to note that adding white noise leads to the occurrence of large zones of non-symmetric beams vibrations under symmetric loads and symmetric boundary conditions applied. Zones of non-symmetric vibrations are shown in Fig. 2c,e.

The carried out analysis of the charts $q_0(\omega_p)$ for $w_n = \{0.0; 0.01; 0.1; 1\}$ have shown that for a lack of the white noise a zone of non-symmetric vibrations is very small and it cannot be easily detected through the applied numerical approaches. However, an occurrence of the white noise of small intensity yields a birth of large zones of non-symmetric vibrations. Further increase of the white noise intensity causes a slight further grow of the non-symmetric zones (see Fig. 3). In the carried out further numerical experiments for two vibrational regimes, i.e. for periodic vibrations (point 1) and chaotic vibrations (point 2) shown in Fig. 3 we have investigated influence on the noise action on the vibrations of the beam for two sets of the beams excitations represented by points 1, 2. In both cases, i.e. for periodic (point 1) and chaotic (point 2) vibrations the beam is subjected to the white noise of intensity $w_n = 0.01$. The snapshots of the bending forms of the beam $w(0 \le x \le 1)$ are reported in Figures 4, 6 for two time instance given in those figures, where solid (dashed) curves correspond to $w_n = 0\,(w_n = 0.01)$. In addition, in Figures 5, 7 the beam center time histories $w(x = 0.5, t)$ associated with the points 1,2 shown in Fig. 3, i.e. for the regular and chaotic beams vibrations, are reported. Dark points correspond to time instants, where the birth of the non-symmetric vibrations have been detected. Remarkably, their number grows in time yielding a collapse of the previous symmetric beam vibrations.

Table 1. The Morlet wavelets, time histories, frequency power spectrum, Poincarè map, 3D phase plot, and beam vibration snapshots in time instant t = 804.145 (point 1)

$w_n = 0, \omega_p = 1.0503, q_0 = 43300$

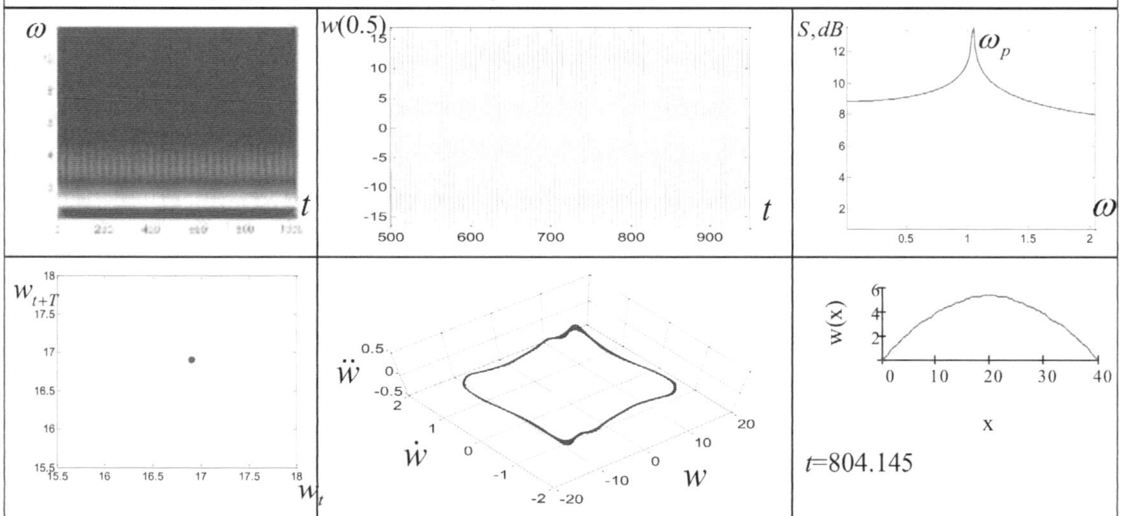

$w_n = 0.01, \omega_p = 1.0503, q_0 = 43300$

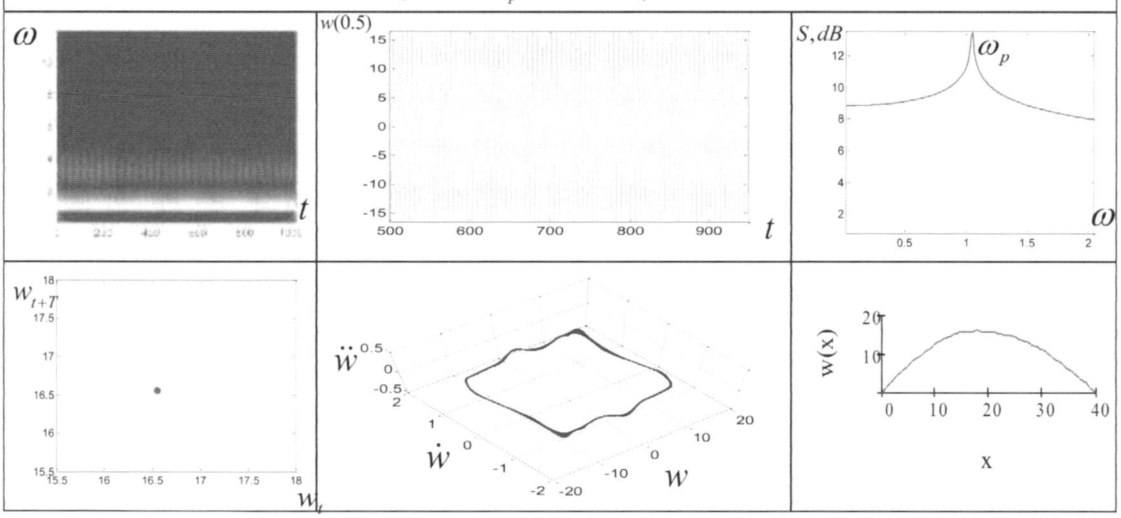

Table 2. The Morlet wavelets, time histories, frequency power spectrum, Poincarè map, 3D phase plot, and beam vibration snapshots in time instant t=804.145 (point 2)

$w_n = 0, \omega_p = 5.5303, q_0 = 56700$

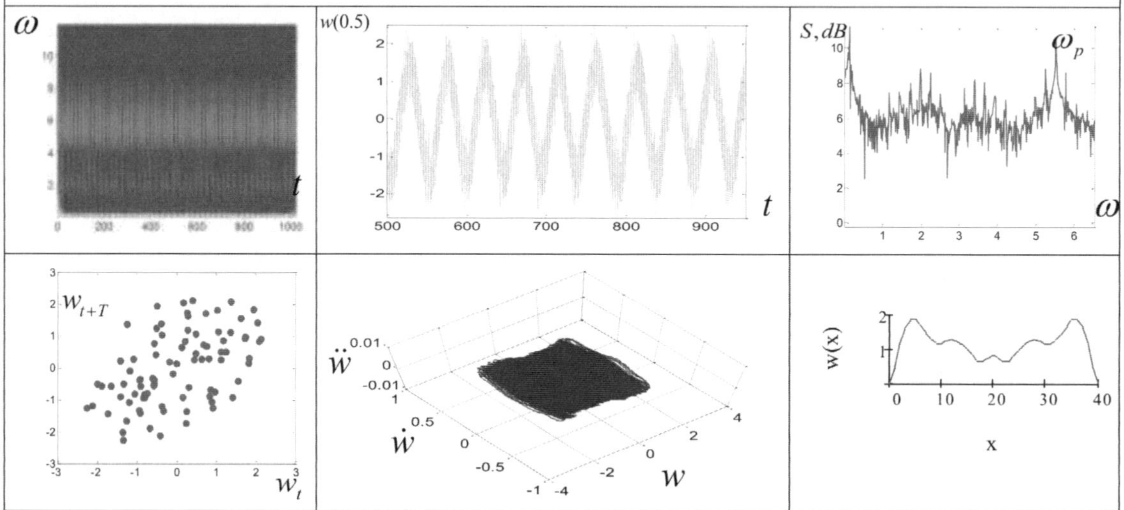

$w_n = 0.01, \omega_p = 5.5303, q_0 = 56700$

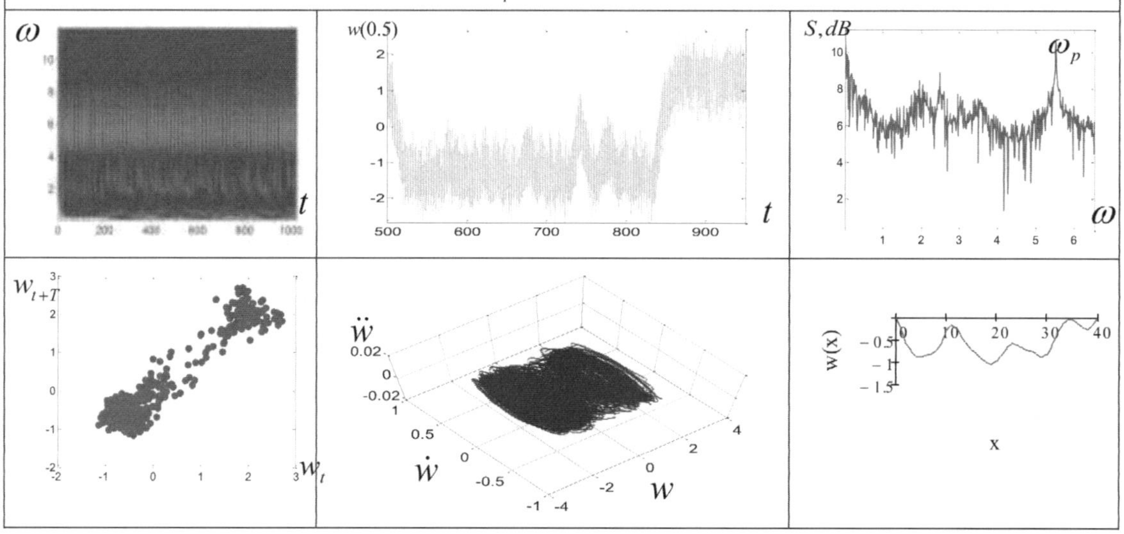

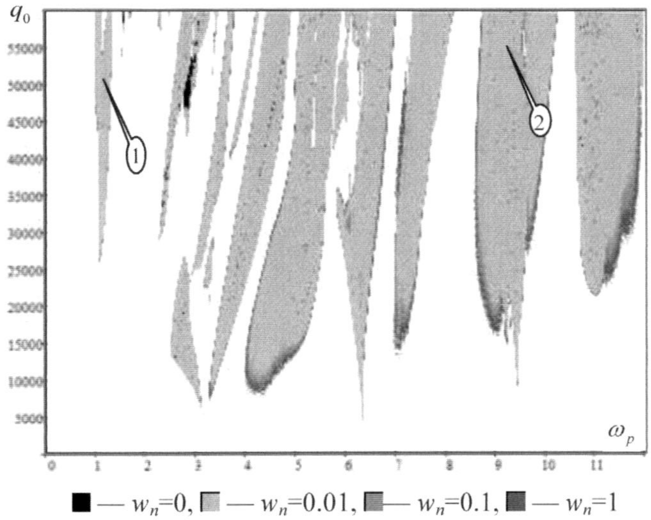

Figure 3. Charts of zones, where non-symmetric beam vibrations appear at least in one instant of time.

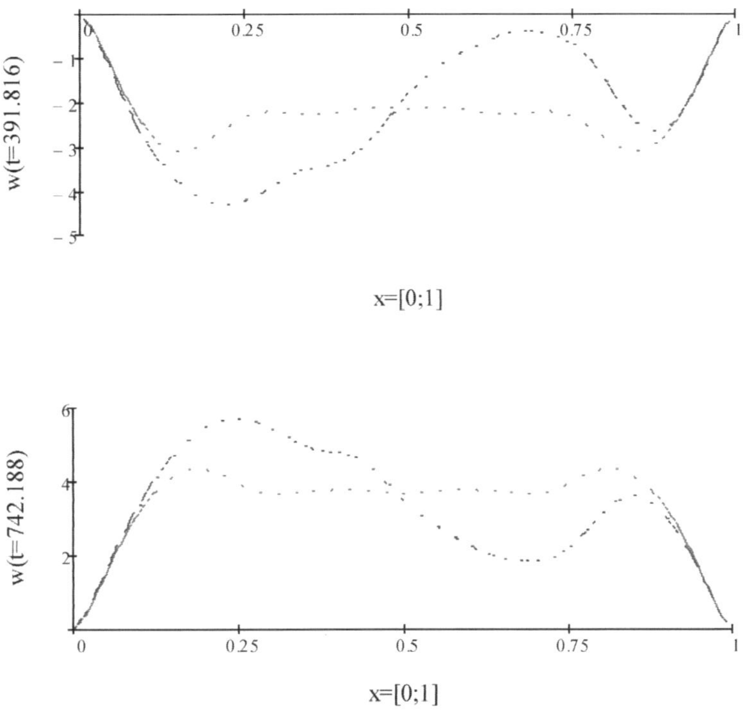

Figure 4. Snapshots of beam center vibrations (point 1) in different time instants.

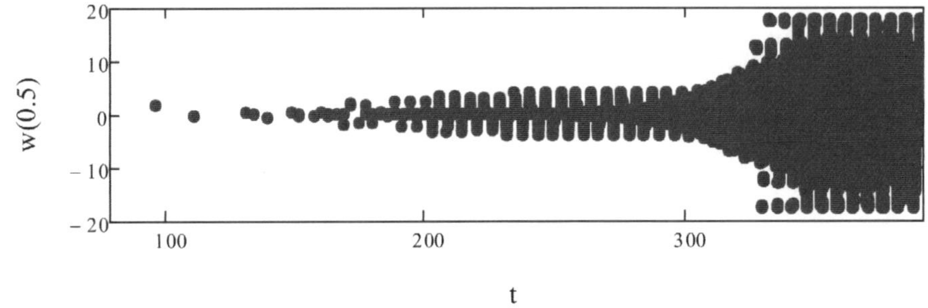

Figure 5. Time history of the beam center vibrations with the marked (in blue) points of symmetry braking (point 1).

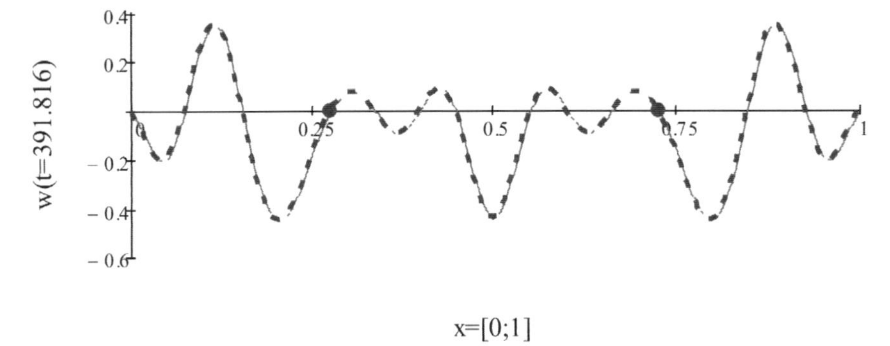

Figure 6. Snapshots of beam center vibrations (point 2) for two different time instants.

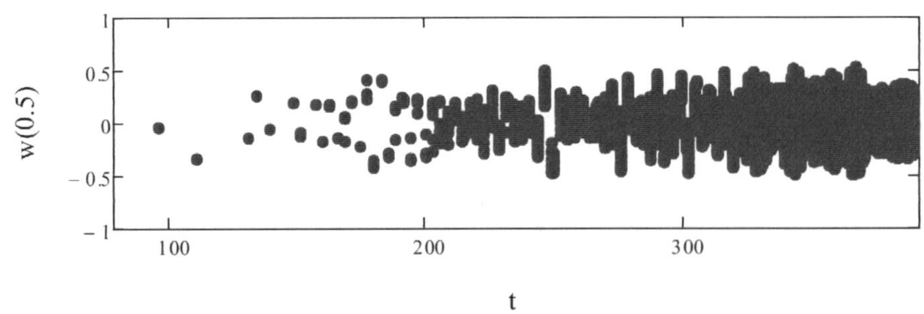

Figure 7. Time history of the beam center vibrations with the marked (in blue) points of symmetry braking (point 2).

Observe that in both points 1, 2 the detected time instants of the non-symmetric vibrations occurrence are approximately the same. In Figure 4 two snapshots of the beam bending forms are reported corresponding to either lack of white noise (solid curve) or action of the white noise intensity $w_n = 0.01$. Figure 6 presents the analogous results for the point 2 (chaotic vibrations), whereas in Figures 5, 7 time histories of the symmetric and non-symmetric beam center vibrations are shown.

Table 3. FFT and Morlet spectra for points 1, 2 with the noise intensity w_n=0, w_n=0.01.

Point 1 ($\omega_p = 1.0853, q_0 = 50000$)			
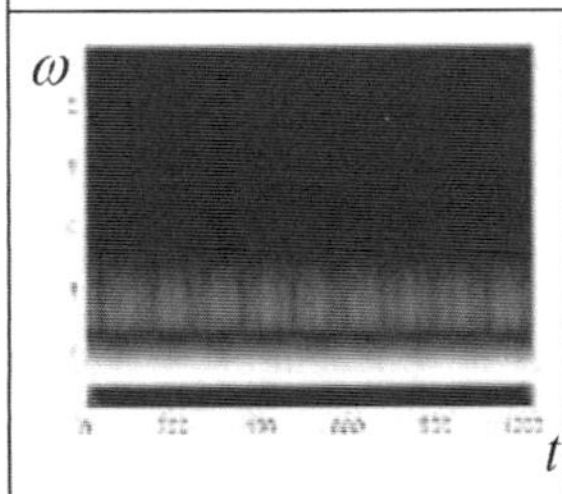	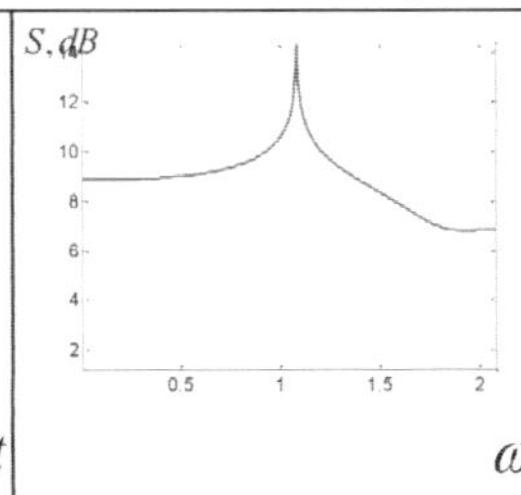	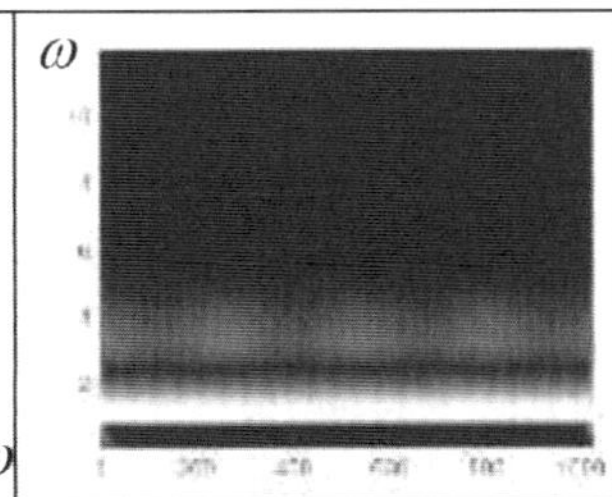	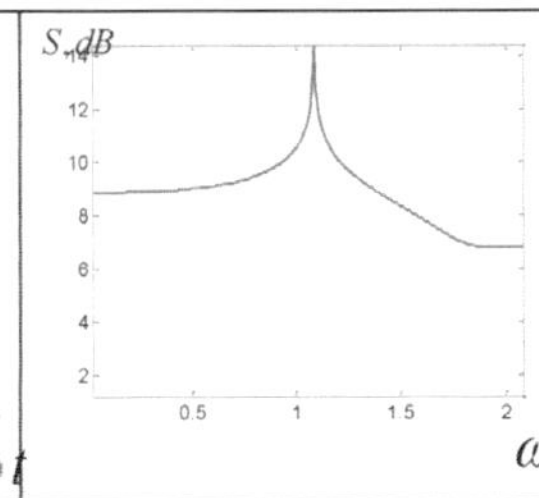
w_n=0		w_n=0.01	
Point 2 ($\omega_p = 9.5903, q_0 = 56400$)			
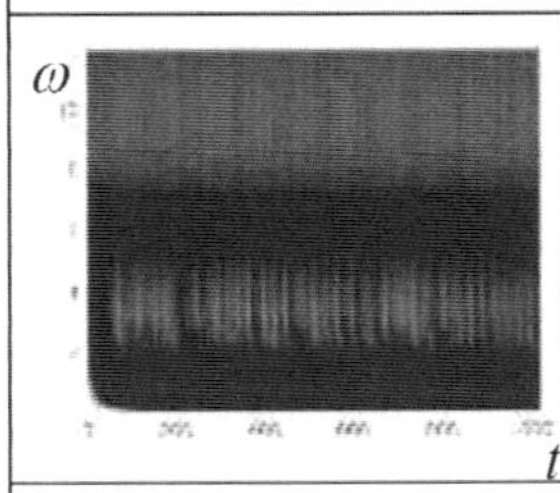	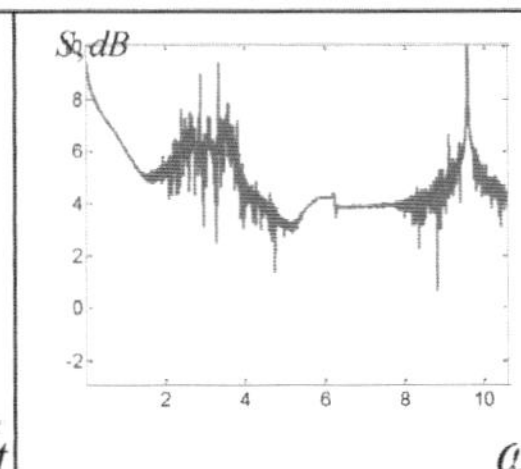	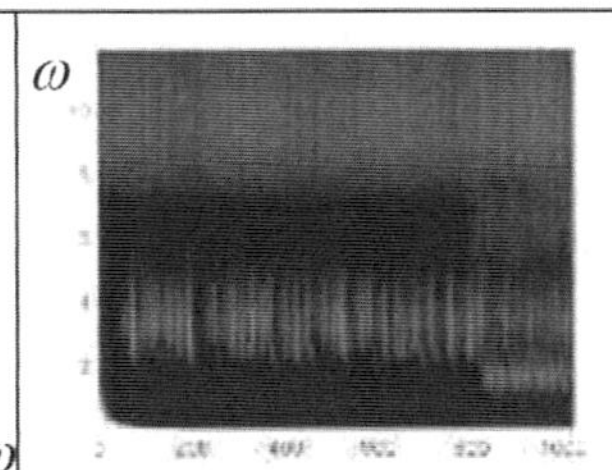	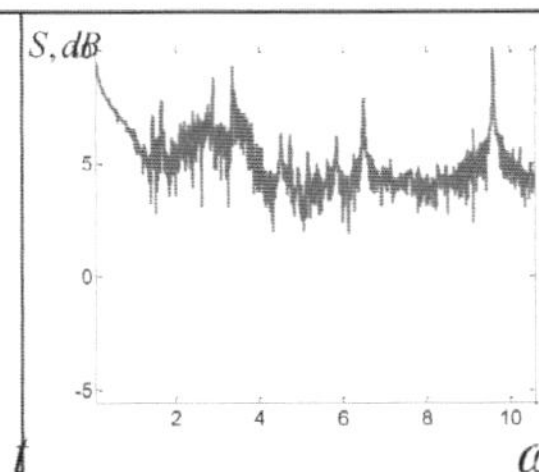
w_n=0		w_n=0.01	

Finally, in Table 3 the Fourier spectrum (FFT) as well as the Morlet wavelet spectrum $\omega(t)$ for the points 1, 2 are given. It should be emphasized that the vibrational regime of both chosen points do not exhibit qualitatively new phenomena. Though in the case of regular beam vibrations (point 1) the noise action does not change the vibration qualitatively, but in the case of point 2 (chaos) the beam chaotic dynamics increases in time, and the chaotic vibrations are transited into hyper-chaotic vibrations.

6. Concluding remarks

We have detected, illustrated and discussed novel non-linear phenomena exhibited by geometrically non-linear Euler-Bernoulli beams under combined harmonic and white noise excitations. In particular, we have reported non-symmetric beam vibrations under symmetrically distributed loads and symmetric boundary conditions applied. Furthermore, the vibration characteristics shown in Tables 1, 2 do not exhibit noisy effects in the case of regular vibrations (point 1), which contradicts the observations of the beam characteristics in the case of point 2 (chaos).

Acknowledgement

This paper is financially supported by the National Science Centre of Poland under the grant MAESTRO 2, No. 2012/04/A/ST8/00738, for the years 2013–2016.

References

[1] Horsthemke W and Lefever R 1984 *Noise-Induced Transitions. Theory and Applications in Physics, Chemistry, and Biology* (Berlin Heidelberg New York Tokyo: Springer-Verlag).

[2] Ruelle D and Takens F 1971 On the nature of turbulence *Comm. Math Phys.* **20** 167-192.
[3] Sprott J C 2010 *Elegant Chaos. Algebraically Simple Chaotic Flows* (Singapore: World Scientific).
[4] Berge P., Dubois M., Manneville P., Pomeau Y. 1980 Intermittency in Rayleigh-Benard convection *J. Phys. Paris Lett.* **41** L341.
[5] Bar-Yoseph P.Z., Fisher D., Gottlieb O. 1996 Spectral element method for nonlinear spatio-temporal dynamics of an Euler-Bernoulli beam *Computational Mechanics* **19** 136-151.
[6] Bokhari A.H., Mahomed F.M., Zaman F.D. 2010 Symmetries and integrability of a fourth-order Euler-Bernoulli beam equation *Journal of Mathematical Physics* **51** 053517.
[7] Dunne J.F., Ghanbari M. 2001 Efficient extreme value prediction for nonlinear beam vibrations using measured random response histories *Nonlinear Dynamics* **24** 71-101.
[8] GeGen, XuJia 2009 Stochastic bifurcation of one flexible beam subject to axial Gauss white noise excitation *IEEE* 1647-1651.
[9] Lan C.B., Qin W.Y. 2014 Energy harvesting from coherent resonance of horizontal vibration of beam excited by vertical base motion *Applied Physics Letters* **105** 113901.
[10] Ling Xu, Qiaozhen Ma 2015 Existence of random attractors for the floating beam equation with strong damping and white noise *Boundary Value Problems* **2015** 126.
[11] Zhang W, Wang F and Yao M 2005 Global bifurcations and chaotic dynamics in nonlinear nonplanar oscillations of a parametrically excited cantilever beam *Nonlinear Dynamics* **40**(3) 251-279.
[12] Awrejcewicz J, Krysko V A and Papkova I V 2012 Routes to chaos in continuous mechanical systems. Part 1. Mathematical models and solution methods *Chaos, Solitons & Fractals* **45** 687-708.
[13] Awrejcewicz J, Krysko V A and Papkova I V 2012 Routes to chaos in continuous mechanical systems. Part 2: Modelling transitions from regular to chaotic dynamics *Chaos Solitons & Fractals* **45** 709-720.
[14] Awrejcewicz J, Krysko V A and Papkova I V 2012 Routes to chaos in continuous mechanical systems. Part 3: The Lyapunov exponents, hyper, hyper-hyper and spatial-temporal chaos, *Chaos Solitons & Fractals* **45** 721-736.
[15] Volmir A S 1972 *Nonlinear Dynamics of Plates and Shells* (Moscow: Nauka), in Russian.

Dynamics behaviour of an elastic non-ideal (NIS) portal frame, including fractional nonlinearities

J M Balthazar[1], R M L F Brasil[2], J L P Felix[3], A M Tusset[4], V Picirillo[4], I Iluik[5], R T Rocha[6], A Nabarrete[1] and C Oliveira[1,7]

[1] ITA-Aeronautics Technological Institute, Mechanical-Aeronautics Division, São José dos Campos, SP, Brazil,
[2] Federal University of ABC, Center for Eng., Modelling and Applied Social Sciences,
Santo Andre, SP, Brazil.
[3] UNIPAMPA - University in Bagé, Alegrete, RS, Brazil.
[4] UTFPR – Federal University of Technology – Paraná, Ponta Grossa, PR, Brazil.
[5] USP, Polytechnic School of the University of Sao Paulo, São Paulo, SP, Brazil,
[6] UNESP – São Paulo State University, Mechanical Engineering, Bauru, SP, Brazil.
[7] UFGD, Federal University of Grande Dourados, Dourados, MS, Brazil.

e-mail :jmbaltha@ita.br

Abstract. This paper overviews recent developments on some problems related to elastic structures, such as portal frames, taking into account the full interactions of the vibrating systems, with an energy source of limited power supply (small motors, electro-mechanical shakers). We include a discussion on fractional (rational) damping and stiffness effects on the adopted modelling. This was a plenary lecture, delivered in the event titled: Mechanics of Slender Structures, organized in Northampton, England from 21-22, September 2015.

1. Introduction

This paper shows a series of problems related to nonlinear dynamics and control, correlated to Electro-mechanical Systems [1] using an elastic support. Here, we analyzed such systems those fall into three groups: the conventional MACRO, MEMS and NEMS electro-mechanical systems, taking into account the existence of a full interaction between mechanical and electrical field quantities. The aim of this paper is to describe a large number of phenomena caused by the action of vibrations on nonlinear electro-mechanical systems and the energy source supply to it. This paper gives a general description of this class of phenomena which has been discussed in recent years.

We remarked that a large number of publications was based on the assumption that the external excitations are produced by an ideal source of power (IS) with prescribed time history: prescribed magnitude, phase and frequency, or in random problems with prescribed characteristics. In reality, the excitations sources are non-ideal and they have always limited power, limited inertia and their frequencies varies according to the instantaneous state of the vibrating system [2].

We also commented that small motors with limited power are used in laboratory tests and therefore the investigations of mutual interaction of driven and driving subsystems are very important. We will discuss several contributions on the past, nowadays and perspective futures in engineering that may be modelled by non-ideal systems (NIS) with many degrees of freedom. In this lecture, we will present examples of this kind of engineering devices restricted to portal frame foundation.

2. Electro mechanical systems

It is well known that the analyses of the motion of real electro-mechanical systems were carried out by means of mathematical models, which have been always simplified and therefore describe the "exact" behavior with some degree of approximation. The study of problems involving the coupling of several systems was widely explored, in the last years, essentially as function of the change of constructive characteristics on machines and structures.

We noted that a lot of oscillatory (vibrating) phenomena of real systems cannot be explained or solved on the basis of a linear theory and it is important to introduce nonlinear characteristics into the mathematical models of vibrating systems and in particular to electro-mechanical systems. The main difficulty for comparisons with linear systems was mainly due to the absence of validity of the superposition principle. Every nonlinear vibrating system should be solved individually and a special methodology must be developed for each class of problems.

Here, we dealt with a special class of nonlinear systems called non-ideal systems (NIS).

By one hand, we verified that the study of the vibrating systems had been considered a major challenge in theoretical and practical engineering research, when the external excitation was influenced by the response of the system. The motion of a vibrating structure under the action of such energy source is accompanied by the interaction between these non-ideal motors and their supports. In usual approaches, the excitation is considered as ideal (IS); that is, the influence of the motion of the structure on the motor is disregarded. Here, the reciprocal influence of the system on the energy source is considered (NIS).

As a direct consequence, in the region of resonance, unstable conditions of motion occur and the form of resonance curve depends on which direction the frequency of the excitation is being altered. We know that the problem of passage through resonance of unbalanced equipment, with an operational speed higher than the lower frequencies of vibration of the supporting structure, has been studied for a long time. However, the machine may not be able to supply the power necessary for this technique, as a large part of its energy is used in shaking the structure and not for increasing the drive speed of the electrical motor. This is well known as Sommerfeld effect (see [2-5], undeserved of others).

Note that as long as the motor drive speed is assumed to be uncoupled from the forced vibrations on the supporting structure, one has a known forcing function of time corresponding to an unlimited or ideal power supply. The introduction of real torque-speed curves for non-ideal motors renders the system nonlinear and capable of multiple steady-state periodic motions whose stability must be assessed. Further complexities may be introduced if the structure itself exhibits nonlinear behavior.

We also found that the right branch of the frequency response curve may be unstable if its negative slope is larger than that of the torque-speed characteristic of the driving motor.

2.1. *Portal frame foundation*

Some new applications related to portal frame (NIS) structure and their nonlinear vibrations have been investigated by several authors [6-9]. The same problem with two non-ideal sources was studied by [10]. The authors considered a description of a non-linear analysis of a metallic portal frame excited by a DC motor. In this referenced work, the two columns of the structure were clamped in their bases to a large seismic mass and the motor was attached to the center of the horizontal beam.

In order to allow the possible observation of a non-linear phenomenon, the dimensions of the structure were chosen in such a way that there was a ratio of one-to-two between the first and second natural frequencies for which the modes are anti-symmetrical or a sway mode and symmetrical or a bending mode. For the same reason, all joints connecting the columns to the horizontal beam were welded, in order to reduce the structural damping. Modal saturation, Sommerfeld effect and other interesting phenomena were observed and demonstrated by using analytical, numerical and experimental techniques.

In [11] the authors examined the nonlinear control method based on the saturation phenomenon and of systems coupled with quadratic nonlinear ties applied to a shear-building portal plane frame foundation that supports an unbalanced direct current with limited power supply (NIS).

The authors in [12] presented some research results on the optimization of the efficiency of an impact damper for a NIS. In the work done by [13] it was analyzed the Sommerfeld effect, where a part of the excited system behaves as energy sink was detected.

In [14] the authors considered both active and passive controls to suppress the chaotic behavior of a simple portal frame, under the excitation of an unbalanced DC motor with limited power supply (NIS). The adopted active control strategy consisted of two controls: the nonlinear feedforward in order to keep the controlled system in a desirable orbit, and the feedback control, which may be obtained by considering the state-dependent Riccati equation to bring the system to the desired orbit, using a magneto rheological (MR) damper. To control the electric current of the MR damper the Bouc-Wen mathematical model was used to the MR damper. The passive control was obtained by means of a nonlinear sub-structure whose properties refers to a nonlinear energy sink. Simulations showed the efficiency of both the passive control (energy pumping) and active control strategies in the suppression of the chaotic behavior.

2.2. *Base excitation effects.*

Besides excitation due to unbalanced rotating machines, several authors analyze vibrating systems subjected to base excitation. It is known that [6] analyzed a frame under support motion considering linear elastic forces. In [15] it was performed numerical and experimental dynamical analysis of a parametrically excited pendulum. In this case the presence of quasi-periodic nonlinear interactions between the structural frame and the electro-mechanical shakers, characters of the non-ideal phenomenon [5, 24,25,26] in the proposed NIS model. We mention that it was carried out a dynamical analysis of a flexible portal frame under base excitation, which is provided by an electro-mechanical shaker in [17]. The electro-mechanical device considered in this work consists of an electric system magnetically coupled to a mechanical structure as used before by [16]. In this paper, the authors investigated the effectiveness of the linear electromechanical vibration absorber (LEVA) and a nonlinear electromechanical vibration absorber (NEVA) in the vibration attenuation for NIS structures. An analysis of the effects of the parameters of coupling and of nonlinear coefficients with increasing of constant torque of the DC motor is presented.

The authors in [18] considered a flexible portal frame parametrically excited by an electro-dynamical shaker. The dynamical jumps, an important point associated with the proposed system were the energy transfer between the shaker and the FPF. This energy transfer is observed when system behaved periodically but it is more significant in quasi-periodic response.

2.3. *Harvesting Energy.*

Recently, much interest has been demonstrated in the concepts of electro-mechanical systems that are able to scavenge or harvest energy from their operating environment. In the process of energy harvesting, the electrical energy is obtained through of conversion of mechanical energy from an ambient vibration source by a type of transductor, as a piezoceramic thin film, which is a material with piezoelectric properties.

In [19] the authors presented a FEM model of a portal frame, which is analyzed by displaying the energy transfer between modes when the structure was excited via the first sway mode. They also showed the saturation of the second mode (vertical symmetrical mode) when the structure was excited by the support motion resonant to this mode, and the energy was transferred to the coupled horizontal sway mode. The computed values were proportional to the power available for energy harvesting in each mode. As a novel feature for this research, we present a situation where the energy introduced in a structural system with a certain configuration was made available for harvesting another vibration mode very different.

In [20] it was presented the extraction of energy from a simple portal frame structure excited via its second (first symmetric) mode. As 2:1 internal resonance is present between that mode and the first (sway) mode, the phenomenon of mode saturation and energy exchange (modal coupling) occurs. Energy pumped into the system through the second (vertical) mode was partially transferred to the horizontal (sway) mode. An evaluation of the energy available for harvesting in each of the considered modes was computed. Here, the nonlinearities present in the piezoelectric material were considered in the piezoelectric coupling mathematical model.

In [21] it was considered the piezoceramic coupled in the column of a portal frame with 2:1 internal resonance may reduce the column vibration and transfer its vibration energy to electrical energy. In this paper we consider a simple portal frame structure with 2:1 internal resonance excited by an electro-dynamical shaker which can simulate a harmonic displacement and a piezoelectric material coupled in a column of the portal frame. We also mention that [22] presented an analysis of a new energy harvester model, based on a (NIS) simple portal frame structure. The horizontal motion of the portal frame was considered under a NIS excitation, and the approximated mathematical model of the system was obtained, considering the coupled oscillators. To model the piezoelectric coupling, the nonlinearities of the piezoelectric material were considered. To check the robustness of the control strategy, an analysis considering uncertainties in the parameters of the model was performed, showing the efficiency of the passive control (energy pumping) in the suppression of the chaotic behavior, as well as the sensitivity of the control system to parametric errors [23]. Also presented a model of energy harvester based on a simple portal frame (NIS). The nonlinearities present in the piezoelectric material are considered in the piezoelectric coupling mathematical model. The system is a bi-stable Duffing oscillator presenting a chaotic behavior. Analyzing the average power variation, and bifurcation diagrams, the value of the control variable that optimizes power or average value that stabilizes the chaotic system in the periodic orbit was determined. The control sensitivity was determined to parametric errors in the damping and stiffness parameters of the portal frame. The proposed passive control technique used a simple pendulum to tune to the vibration of the structure to improve the energy harvesting.

2.4. *NIS Emergent problems.*

In the papers [24] the authors analyzed a proposed mathematical model, by using numerical integrations, taking into account the reciprocal influence between the vertically excitation, due an electro-dynamical shaker (EDS) and the vibration response of a tuning fork beam (TFB).The internal physical components of a shaker consist of an electric component coupled magnetically to a mechanical structural one. The behavior of the shaker type utilized here is similar of the one used by [15] also [27] studied a tuning fork vibratory micro machined gyroscope subjected to base excitation. Here, the investigation was based the behavior for the macro system. The (TFB) is modeled by two inverted pendulums of motion in the opposite directions hung by the same rods of vertical and horizontal motion according to suggestions by [28,29].Recently, the researcher in [29] investigate the interaction of the dynamics of the electro-shaker with the gyroscope, demonstrated that under certain parameters the system can exhibit complex dynamic behavior such as chaotic motion [30], analyze the influence on the electrical charge, between the cantilever beam and the wide electrode by using numerical methods. For the analysis, the authors used the mathematical model of a Duffing oscillator, under electrostatic effects, with variable capacitor, investigating the dynamic interaction, between a micro machined rate gyroscope and variable force actuators.

3. Rational nonlinearities and numerical simulations results

By other hand, Fractional stiffness and damping is appearing in different contexts in any systems with memory and hysteresis. Such damping is defined by a fractional derivative in contrary to classical viscous damping term with the first order derivative. As the memory of the dynamical system induces extra degree of freedom for the phase space the standard methods of dynamical response analysis and system identification, which relies on the knowledge of system dimensionality cannot be used. Here, we present nowadays results in order to start new researches on the NIS portal frame. The results present here is an extension of the paper [31] using an uncontrolled model of portal frame (Figure 1).

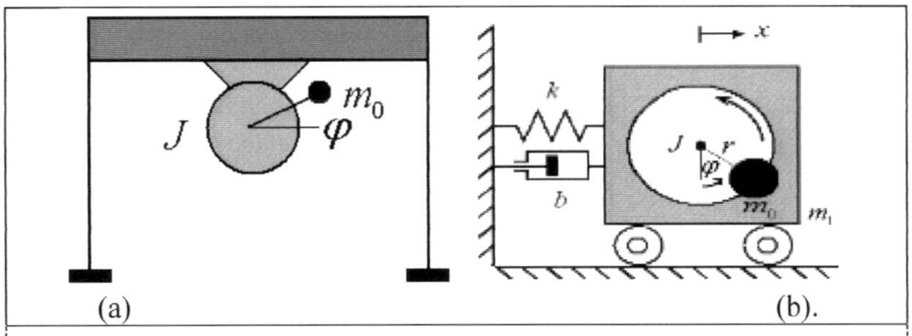

Figure 1.. (a) Non-ideal Portal Frame. (b) Schematic of a non-ideal oscillator **(NIS)**

For this work, the parameters of the coupled dynamical system consist of ($m_0, m_1, k_1, k_{nl}, c_1, x_1, \varphi, J, r, d, s$), the mass, unbalanced mass, linear stiffness, non-linear stiffness, linear damping, displacement, angular displacement, inertia moment, eccentricity of the unbalanced mass, d is related to the voltage applied across the armature of the DC motor and s is a constant for each model of DC motor considered. The resulting mathematical model of the structure is a Duffing-like equation:

$$\begin{aligned}&(m_1+m_0)\ddot{x}+b\dot{x}-k_l x+k_{nl}x^3=m_0 r\left(\ddot{\varphi}\sin(\varphi)+\dot{\varphi}^2\cos(\varphi)\right)\\&(J+r^2 m_0)\ddot{\varphi}-rm_0\ddot{x}\sin(\varphi)=L(\dot{\varphi})=d-s\dot{\varphi}\end{aligned}\qquad (1)$$

Next, we render Eqs. (1) dimensionless in terms of new variables defined by: $\tau=\omega t$, $x_1=\frac{x}{x^*}$, *and* $x_3=\frac{\varphi}{\varphi^*}$, where x^* and φ^* are constant characteristics. Equations (1) can be represented in a state space, in dimensionless form, as:

$$\begin{aligned}&x_1'=x_2\\&x_2'=-\alpha x_2+\beta_1 x_1-\beta_3 x_1^{\,3}+\delta_1\sin(\varphi^* x_3)x_4'+\delta_1\cos(\varphi^* x_3)x_4^{\,2}\\&x_3'=x_4\\&x_4'=\rho_1\sin(x_3)x_2'-\rho_3 x_4+\rho_2\end{aligned}\qquad (2)$$

where: $\alpha\frac{b}{(m_1+m_0)\omega}$, $\omega = (k_1/(m_1+m_0))^{1/2}$, $\beta_1\frac{k_l}{(m_1+m_0)\omega^2}$, $\beta_3=\frac{k_{nl}x^{*2}}{(m_1+m_0)\omega^2}$,

$\delta_1=\frac{m_0 r\varphi^*}{(m_1+m_0)x^*}$, $\rho_2=\frac{d}{(J+r^2m_0)\omega^2\varphi^*}$, $\delta_2=\frac{m_0 r\varphi^{*2}}{(m_1+m_0)x^*}$, $\rho_1=\frac{rm_0x^*}{(J+r^2m_0)\varphi^*}$, $\rho_3=\frac{s\omega\varphi^*}{(J+r^2m_0)\omega^2\varphi^*}$.

3.1. *Dynamic analysis of a fractional-order*

Differential equations may involve Riemann–Liouville differential operators of fractional-order $q>0$, which generally take the form below [32], [33]:

$$D^q x(t)=\frac{1}{\Gamma(\eta-q)}\int_{t_0}^{t}\frac{x^{(\eta)}(u)}{(t-u)^{q-\eta+1}}du \tag{3}$$

Where η is the first integer not less than q. It is easily proved that the definition is the usual derivatives definition when $q=1$. The case $0<q<1$ seems to be particularly important. For simplicity and without loss of generality, in the following, we assume that: $t_0=0,\ 0<q<1$.

Considering the force of dissipation $(-\alpha x_2)$ are fractional order and considering $x_2=x_6$ *as* equation:

$$\begin{aligned}
&x_1'=x_2\\
&x_2'=-\alpha x_6+\beta_1x_1-\beta_3x_1^3+\delta_1\sin(\varphi^*x_3)x_4'+\delta_1\cos(\varphi^*x_3)x_4^2\\
&x_3'=x_4\\
&x_4'=\rho_1\sin(x_3)x_2'-\rho_3x_4+\rho_2\\
&x^{q_1}{}_5=x_6\\
&x^{q_2}{}_6=-\alpha x_6+\beta_1x_5-\beta_3x_5^3+\delta_1\sin(\varphi^*x_3)x_4'+\delta_1\cos(\varphi^*x_3)x_4^2
\end{aligned} \tag{4}$$

Consider the numerical values: $\alpha_1=0.1$, $\beta_1=1$, $\beta_3=2$, $\delta_1=8.373$, $\rho_1=0.05$, $\rho_2=100$, $\rho_3=200$, $x_1(0)=0$, $x_2(0)=0$, $x_3(0)=0$, $x_4(0)=0$, $x_5(0)=0$, $x_6(0)=0$. As the force dissipation depends $x_2=x_6$ it will be considered $q_1=1$ and $0.5\le q_2\ge 1$.

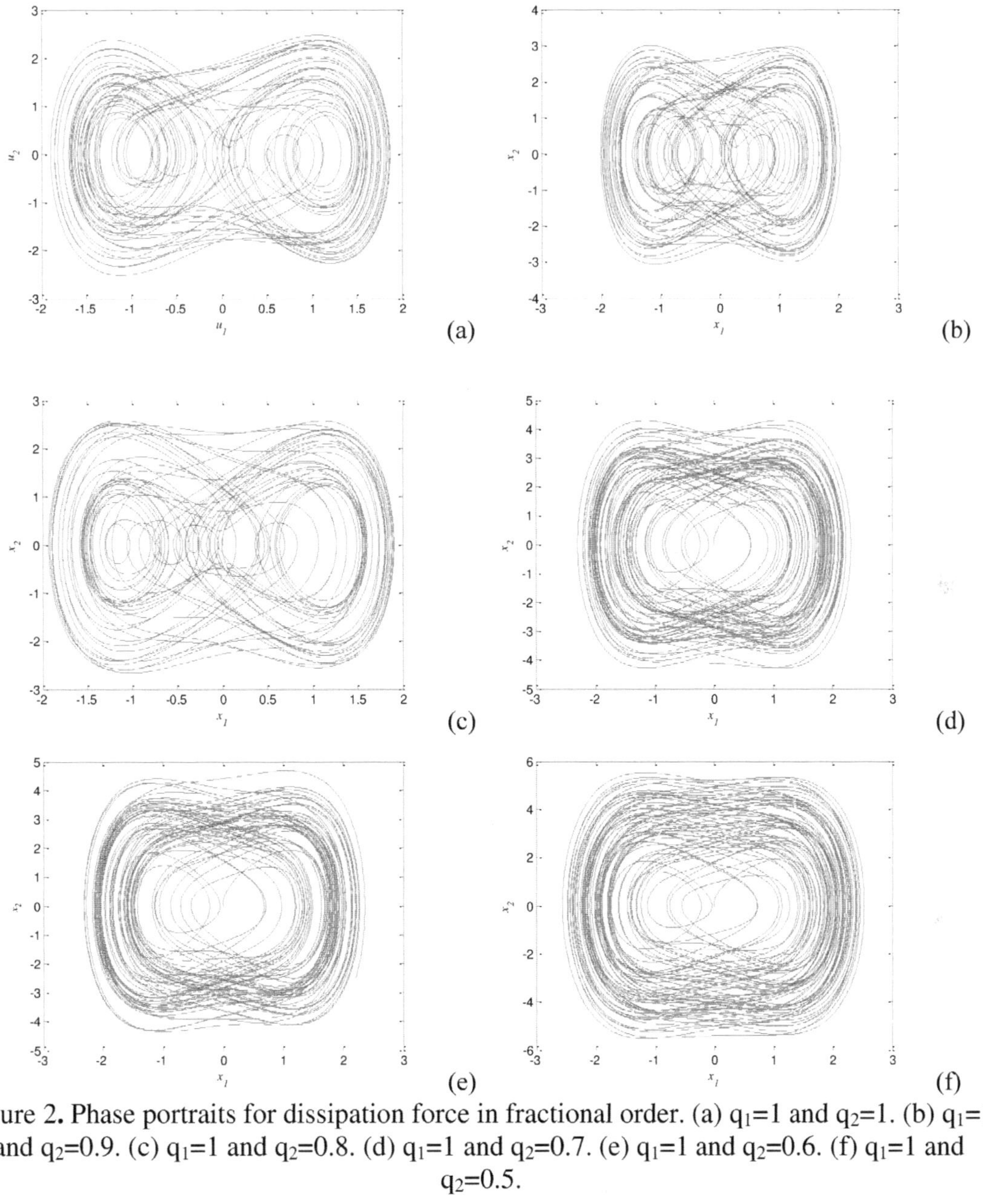

Figure 2. Phase portraits for dissipation force in fractional order. (a) q_1=1 and q_2=1. (b) q_1=1 and q_2=0.9. (c) q_1=1 and q_2=0.8. (d) q_1=1 and q_2=0.7. (e) q_1=1 and q_2=0.6. (f) q_1=1 and q_2=0.5.

As can be seen in Figure 2, the power dissipation is of the order of fractional order derivative will influence the damping. As can be seen in figure 2 the variation of the derivative order generated increase in displacement x_1 and x_2. Since for q_1=1 and q_2=0.5 (figure 2f) was the largest displacement amplitude, indicating reduction of power dissipation.

We can also observe that the variation q_2 influence the behavior of the chaotic system. To check which variation of chaotic behavior is considered the 0-1 test. The 0-1 test for chaos takes as input a time series of measurements and returns a single scalar value of either 0 for periodic attractors or 1 for chaotic attractors [31,34]. According to [35] value of (K_c) can be obtained from:

$$k_c = \frac{cov(X,M(c))}{\left(var(X)var(M(c))\right)^{1/2}} \tag{5}$$

where vectors $X=[1,2,\ldots,n_{max}]$, and $M(c)=[M(1,c),\ldots, M(1, n_{max})]$, $c \in (0,\pi)$ is a fixed frequency chosen arbitrarily, and:

$$M(n,c) = \lim_{n\to\infty} \frac{1}{N}\sum_{j=1}^{N}\left[\left(p(j+n)-p(j)\right)^2 + \left(q(j+n)-q(j)\right)^2\right] \tag{6}$$

$$p(i) = \sum_{j=0}^{i} \frac{(x_j-\bar{x})}{\sigma_x}\cos(jc) \tag{7}$$

$$q(i) = \sum_{j=0}^{i} \frac{(x_j-\bar{x})}{\sigma_x}\sin(jc) \tag{8}$$

where: $\bar{x}$ and σ_x is the mean value and square deviation of examined x_i series, and N is the length of the sampled points in the displacement time series.

For a value of $K_c \cong 0$ indicates a non-chaotic data set while a value of $K_c \cong 1$ indicates a chaotic data set. In table 1 shows the K_c variation for the derivate order $q_1 = 1$ and $0.5 \le q_2 \ge 1$.

Table1: Variation of the chaotic behavior for derivative order variation

q1	q2	Kc
1	1	0.9427
1	0.9	0.8571
1	0.8	0.9038
1	0.7	0.9277
1	0.6	0.8453
1	0.5	0.9562

As can be seen in Table 1, the variation q_2 changes the chaotic behaviour of the system. Reducing for $1 < q_2 \le 0.6$ and increasing to $q_1 = 0.5$. Demonstrating the influence of the order of the derivative ($x^{q}-$) in the system damping.

Considering the force of stiffness and dissipation $(-\alpha x_2 + \beta_1 x_1 - \beta_3 x_1^{\,3})$ are fractional order and considering $x_1 = x_5$ *and* $x_2 = x_6$ as equation:

$$\begin{aligned}
x_1' &= x_2 \\
x_2' &= -\alpha x_6 + \beta_1 x_5 - \beta_3 x_5^{\,3} + \delta_1 \sin(\varphi^* x_3) x_4' + \delta_1 \cos(\varphi^* x_3) x_4^{\,2} \\
x_3' &= x_4 \\
x_4' &= \rho_1 \sin(x_3) x_2' - \rho_3 x_4 + \rho_2 \\
x^{q_1}{}_5 &= x_6 \\
x^{q_2}{}_6 &= -\alpha x_6 + \beta_1 x_5 - \beta_3 x_5^{\,3} + \delta_1 \sin(\varphi^* x_3) x_4' + \delta_1 \cos(\varphi^* x_3) x_4^{\,2}
\end{aligned} \tag{9}$$

Consider the numerical values: $\alpha_1 = 0.1$, $\beta_1 = 1$, $\beta_3 = 2$, $\delta_1 = 8.373$, $\rho_1 = 0.05$, $\rho_2 = 100$, $\rho_3 = 200$, $x_1(0) = 0$, $x_2(0) = 0$, $x_3(0) = 0$, $x_4(0) = 0$, $x_5(0) = 0$, $x_6(0) = 0$, $q_1 = q_2$, and $0.5 \le q_2 \le 1$.

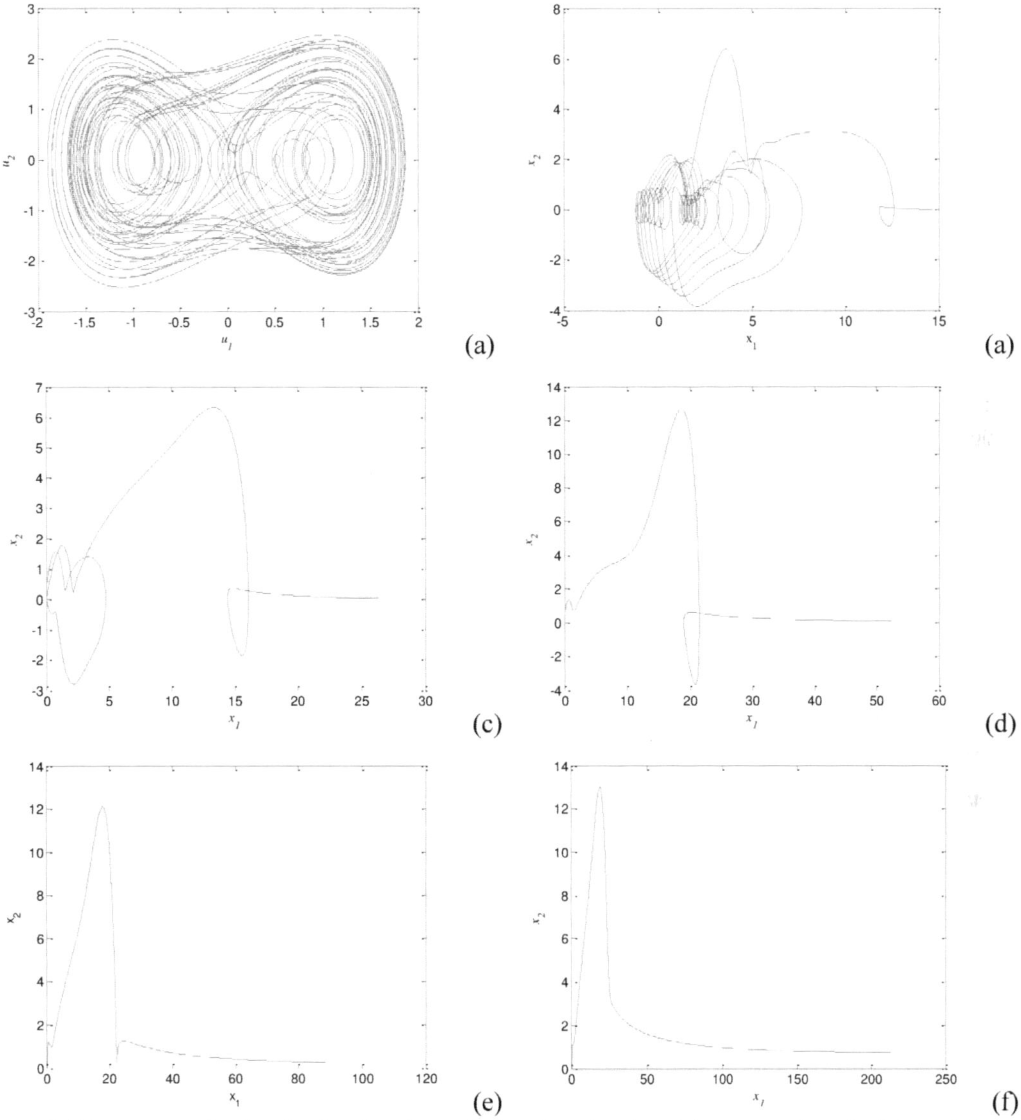

Figure 3. Phase portraits for dissipation force in fractional order. (a) $q_1=1$ and $q_2=1$. (b) $q_1=0.9$ and $q_2=0.9$. (c) $q_1=0.8$ and $q_2=0.8$. (d) $q_1=0.7$ and $q_2=0.7$. (e) $q_1=0.6$ and $q_2=0.6$. (f) $q_1=0.5$ and $q_2=0.5$.

As can be seen in figure 3, in the case of fractional-order stiffness the displacement of the structure amplifies so as to compromise the physical integrity of the structure, whereas the amplitude increases as the order reduced. The results show that for damping in fractional order, the displacement of the structure is chaotic, while in case of the structure stiffness fractional order that the displacement of the structure is amplified according is reduced to the order of the derivative (q_).

4. Conclusions

We presented an analysis of a dynamical coupling between energy sources and structural response that must not be ignored in real engineering problems, since real motors have limited output power, taking into account a portal frame structure support. The used NIS mathematical model was represented with two coupled nonlinear differential equations, including rational nonlinearities.

The main results obtained showed that: As can be seen in figure 3, in the case of fractional-order stiffness is the system becomes unstable, whereas the amplitude increases as the order q_2 reduced. And the damping is the fractional change order chaotic behavior of the system, but keep the system stable, while in the case of the system stiffness is unstable. Futures works will consider control strategies in order to overcome the main problems appoint here as well an investigation of the dynamics, including chaos, periodic motions, multiperiodic and quasiperiodic motions and an analysis of basins of attraction, considering the dynamical integrity [36].

References

[1] Bishop R H 2002 *The Mechatronics Handbook* (CRC Press LLC)

[2] Kononenko V O 1969 *Vibrating System of Limited Power Supply* (ILife).

[3] Balthazar J M, Mook D, Weber H I, Brasil R M L R F, Fenili A, Belato D, Felix J L P 2003 An overview on non-ideal vibrations *Meccanica* **38(2)** 613-21.

[4] Dantas M J H and Balthazar J M 2007 On the existence and stability of periodic orbits in non-ideal problems: general results, ZAMM- *Journal of Applied Mathematics and Mechanics*,**58**, 940-56.

[5] Felix J L P, Balthazar J M and Brasil R M L R F 2009 Comments on nonlinear dynamics of a non-ideal Duffing-Rayleigh oscillator: Numerical and analytical approaches, *Journal of Sound and Vibration.***3(23)**, 1136–49

[6] Barr A D S and McWannell D C 1971 Parametric instability in structures under support motion, *Journal of Sound and Vibration* **14(4)** 491–509

[7] Simitses G J, Swisshelm J D and Vlahinos A S 1984 Flexibly-jointed Unbraced Portal Frames, *J. Construct. Steel Research*, **4** 27-44.

[8] Mazzilli C E N and Brasil R M L R F 1995 Effect of Static Loading on the Nonlinear Vibrations of a Three-Time Redundant Portal Frame: Analytical and Numerical Studies, *Nonlinear Dynamics*, **8** 347-66.

[9] Brasil R M L R F 1999 Multiple Scales Analysis of Nonlinear Oscillations of a Portal Frame Fondations for Several Machines, *Journal of the Brazilian Society of Mechanical Sciences*, **21 (4)** 641-54.

[10] Balthazar J M, Brasil R M L R F and Garzeri F J 2004 On non-ideal simple portal frame structural model: experimental results under non-ideal excitation, *Applied Mechanics and Materials*, **1-2** 51-58.

[11] Felix J L P, Balthazar J M and Brasil R M L R F 2005 On saturation control of a non-ideal vibrating portal frame foundation type shear-building, *Journal of Vibration and Control* **11(1)** 121-36.

[12] Silva M A and Brasil R M L R F 2007 Optimization of impact damping for a nonideal structural system. *CMNE/CILAMCE 2007 Porto*, 13-15 Junho, APMTAC, Portugal.CD ROM

[13] Felix J L P, Balthazar J M and Dantas M J H 2009 On energy pumping,synchronization and beat phenomenonin a nonideal structure coupled to an essentially nonlinear oscillator, *Nonlinear Dynamics* **56(1-2)** 1-11

[14] Tusset A M, Balthazar J M and Felix J L P 2012 On elimination of chaotic behavior in a non-ideal portal frame structural system, using both passive and active controls, *Journal of Vibration and Control*,**19(6)** 803-813.

[15] Xu X, Pavlovskaia E, Wiercigroch M, Romeo F and Lenci S 2007 Dynamic interactions between parametric pendulum and electro-dynamical shake, *Z. Angew. Math. Mech.*, **87(2)** 172 – 186.

[16] Felix J L P and Balthazar J M 2008 Comments on a Nonlinear and Nonideal Electromechanical Damping Vibration Absorber, Sommerfeld Effect and Energy Transfer, *Nonlinear Dynamics*

[17] De Paula A S, Balthazar J M and Felix J L P 2012 Nonlinear Dynamics of a Flexible Portal Frame under Support Excitation, *Procedings of the ICNPAA - World CongressMathematical Problems in Engineering, Aerospace and Sciences*, Viena,AIP Conference Proceedings, **1493** 957-963.

[18] De Paula A S, Balthazar J M and Felix J L P 2013 Some comments on the nonlinear dynamics of a portal frame under basis excitation, *Shock and Vibration* **20 (6)**1093-1113

[19] Iliuk I, Brasil R M L R F , Balthazar J M, Tusset A M, Piccirillo V and Piqueira J R C 2014 Potential application in energy harvesting of intermodal exchange in a frame: *FEM analysis, International Journal of Structural Stability and Dynamics*, **14 (08)**

[20] Balthazar J M, Rocha R T,Tusset A M, Brasil R M L R F, De Pontes Jr B R, Silveira M, Bueno A M and Felix J L P 2014 On Mode Saturation phenomenon as an energy harvester generator. *Proceedings of ENOC 2014 8th European Nonlinear Oscillations Conferences.*July **6-8**, Vienna, Austria.

[21] Rocha R T, Balthazar J M, Tusset A M, Piccirillo V, Felix J L P and Brasil R M L R F 2015 On Energy harvesting of a flexible portal frame support exploiting the saturation phenomenon. *Proceedings of IDETC/CIE 2015. ASME 2015 International Design Engineering Technical Conferences &Computers and Information in Engineering Conference,* August 2-5, Boston, Massachusetts, USA

[22] Iliuk I, Balthazar J M, Tusset A M, Piqueira J R C, De Pontes B R and Felix J L P 2013 A non-ideal portal frame energy harvester controlled using a pendulum, *The European Physical Journal Special Topics* **222** (**7**),1575-86

[23] Iliuk I, Balthazar J M,Tusset A M, Piqueira J C R, Pontes B R, Felix J L P and Bueno A M 2013 Application of passive control to energy harvester efficiency using a nonideal portal frame structural support system, *Journal of Intelligent Material Systems and Structures*, 045389X13500570.

[24] Balthazar J M, Felix J L P and Brasil R M R F 2010 On an energy transfer and nonlinear, non ideal and chaotic behavior of a macro tuning fork beam(tfb), under an electro-dynamical shaker excitation. *Proceedings of 11th Pan-American Congress of Applied Mechanics (PACAM) 2010*,January 04-08, Foz do Iguaçu, PR, Brazil,CD ROM.

[25] Awrejcewicz J, Starosta R and Sypniewska-Kaminska G 2015 Decomposition of governing equations in the analysis of resonant response of a nonlinear and non-ideal vibrating system, *Nonlinear Dyn.* **82** 299–309

[26] Awrejcewicz J, Starosta R and Sypniewska-Kaminska G 2014 Decomposition of the Equations of Motion in the Analysis of Dynamics of a 3-DOF Nonideal System, *Mathematical Problems in Engineering* 8 pages.

[27] Jiang T, Wang A and Liu G 2006 Detection capacitance analysis method for tuning fork micromachined gyroscope based on elastic body model, *Sensors and Actuators A*,**128** 52-59.

[28] Lee Y, Frank P and Feng Z C 2008 Nonlinear complex response of a parametrically excited tuning fork, *Mechanical Systems and Signal Processing,* **22(5)**, 1146-56.

[29] Tehrani M G, Balthazar J M and Silveira M 2015 Parametric study of a Macro-Scale Tuning fork gyroscope. International Conference on Engineering Vibration, Ljubljana, Slovenia, September 7-10

[30] Chavarette F R, Balthazar J M, Felix J L P 2013 On Nonlinear Dynamics and Control Design in a MEMS. Dynamics for advances tecnologies and engineering design, IUTAM book series, 404-419

[31] Tusset A M, Piccirillo V, Balthazar J M and Brasil R M R F 2015 On Suppression of chaotic motions of a portal frame structure under non-ideal loading using a magneto-rheological damper. *Journal of Theoretical and Applied Mechanics* ,**53** **(3)** 653-664

[32] Yu Y, Li H X, Wang S and Yu J 2009 Dynamic analysis of a fractional-order Lorenz chaotic system. *Chaos, Solitons and Fractals* **42**, 1181-89

[33] Tusset A M, Balthazar J M, Bassinello D, Pontes JR B R and Felix J L P 2012 Statements on chaos control designs, including a fractional order dynamical system, applied to a MEMS combdrive actuator. *Nonlinear Dynamics* **69**, 1837-5

[34] Felix J L P, Silva E L, Balthazar J M, Tusset A M, Bueno A M and Brasil R M L R F 2014 On nonlinear dynamics and control of a robotic arm with chaos. *MATEC Web of Conferences*, **16**, p. 05002

[35] Gopal R, Venkatesan A and Lakshmanan M 2013 Applicability of 0-1 test for strange nonchaotic attractors, *Chaos* **23**, 023123-1:15

[36] Rega G and Lenci S 2008 Dynamical Integrity and Control of Nonlinear Mechanical Oscillators, *Journal of Vibration and Control*, **14(1-2)**:159-179

Modal analysis of sailplane and transport aircraft wings using the dynamic stiffness method

J R Banerjee

School of Mathematics, Computer Science and Engineering City University London, Northampton Square, London EC1V 0HB, UK

Abstract. The purpose of this paper is to provide theory, results, discussion and conclusions arising from an in-depth investigation on the modal behaviour of high aspect ratio aircraft wings. The illustrative examples chosen are representative of sailplane and transport airliner wings. To achieve this objective, the dynamic stiffness method of modal analysis is used. The wing is represented by a series of dynamic stiffness elements of bending-torsion coupled beams which are assembled to form the overall dynamic stiffness matrix of the complete wing. With cantilever boundary condition applied at the root, the eigenvalue problem is formulated and finally solved with the help of the Wittrick-Williams algorithm to yield the eigenvalues and eigenmodes which are essentially the natural frequencies and mode shapes of the wing. Results for wings of two sailplanes and four transport aircraft are discussed and finally some conclusions are drawn.

1. Introduction

Sailplane and transport aircraft wings are slender and flexible because of their high aspect ratios resulting from large spans and relatively short chords. As a consequence, they are easily prone to vibration problems. In this respect, modal analysis of aircraft wings, particularly those with high aspect ratios is very important. Sailplane and transport airliner wings are typical examples for which the investigation is of great significance. Indeed modal analysis plays an important role in the design of aircraft wings. An analysis of this kind is an obligatory airworthiness requirement which is rigorously enforced by the civil aviation authorities. The purpose of this paper is to carry out such an analysis and investigate the modal behaviour of sailplane and transport aircraft wings by applying the dynamic stiffness method.

One of the main motivations for modal analysis of aircraft wings originates from that fact that it is a fundamental prerequisite to carry out an aeroelastic or response analysis, particularly when using the normal mode method. There are some published papers in this and related areas [1-7]. In general, the finite element method (FEM) is widely used to investigate the modal behaviour of aircraft wings. The FEM is an approximate method where the stiffness and mass properties of all individual elements are assembled to form the overall stiffness matrix **[K]** and mass matrix **[M]** of the structure which is an aircraft wing here. Then for modal analysis, upon imposing the boundary conditions, the typical eigenvalue problem of the type $[[\mathbf{K}] - \lambda[\mathbf{M}]]\{\Delta\} = \mathbf{0}$ is solved where $\{\Delta\}$ is the nodal displacement vector and the square root of λ gives the natural frequencies of the structure. The corresponding mode shapes are recovered in the usual way. In the FEM it is generally true that by increasing the number of elements in the analysis, the results become more and more accurate. It is acknowledged that the FEM is numerically intensive and the degrees of freedom identified by the order of **[K]** and **[M]** matrices decide the number of eigenvalues (which are essentially the natural frequencies) that can be computed.

The higher order natural frequencies will of course, be considerably less accurate. Against this background, there is an elegant and powerful alternative to the FEM for modal analysis of aircraft wings or any other structures. This method is the so-called dynamic stiffness method (DSM). The DSM unlike the FEM, relies on an exact single frequency dependent dynamic stiffness element containing both the mass and stiffness properties of the element as the basic building block. The assembly procedure in the DSM is essentially the same as it is in the FEM, but a single dynamic stiffness element matrix is used for each structural component instead of separate mass and stiffness matrices to form the overall frequency-dependent dynamic stiffness matrix $[\mathbf{K_D}]$ of the complete structure (wing). The eigenvalue problem is formulated as $[\mathbf{K_D}]\{\Delta\}=0$ where $\{\Delta\}$ is the nodal displacement vector comprising the amplitudes of nodal displacements. The next step is to extract the eigenvalues of the structure. At this point a significant difference with the FEM arises with regard to the solution technique. The formulation $[\mathbf{K_D}]\{\Delta\}=0$ leads to a transcendental (nonlinear) eigenvalue problem as opposed to the linear eigenvalue problem generally encountered in the FEM. The best available solution technique to extract the eigenvalues in the DSM is to apply the algorithm of Wittrick and Williams [8], known as the Wittrick-Williams algorithm in the literature which has featured in literally hundreds of papers. The algorithm which monitors the Sturm sequence property of the dynamic stiffness matrix is robust and it ensures that no natural frequency of the structure is missed.

Within the above context, a range of aircraft wings is investigated for their free vibration characteristics in this paper. Two different categories of aircraft wings are analysed. They are essentially for sailplane and transport airliner wings. Two illustrative examples for the former and four for the latter are demonstrated when presenting numerical results. The investigation required considerable efforts for data preparation to model each of the wings. The dynamic stiffness method which provides the best possible model accuracy is used as mentioned. In idealising the wing, an assembly of the frequency dependent dynamic stiffness elements of bending-torsion coupled beams [9-10], comprising both the mass and stiffness properties is efficiently utilised. Natural frequencies and mode shapes computed from the dynamic stiffness method are compared and contrasted and finally some conclusions are drawn.

2. Theory

2.1 Dynamic stiffness matrix of a bending-torsion coupled beam

An aircraft wing such as the one shown in figure 1 is a classic example of a bending-torsion coupled beam. Such a representation is particularly relevant to analyse a high aspect ratio wing.

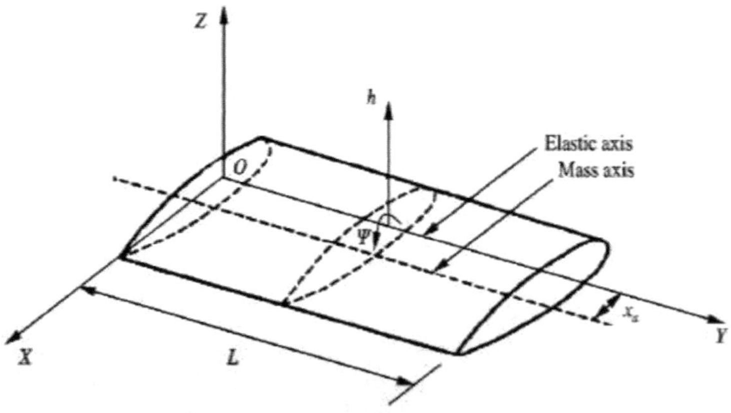

Figure 1 An aircraft wing idealised as a bending-torsion coupled beam.

In essence, the coupling between the bending and the torsional motions arises due to non-coincident mass and elastic axes which are respectively the loci of the centroid and shear centres of the beam cross-section. Thus for an aircraft wing it is not generally possible to realize a torsion-free bending displacement or a bending-free torsional rotation during its dynamic motion unless the load or the torque is applied through or about the shear centre. Given this perspective, a high aspect ratio non-uniform aircraft wing can be accordingly modelled as an assemblage of bending-torsion couple beams of the type shown in figure 1. This paper uses a dynamic stiffness approach and develops the dynamic stiffness matrix of a uniform bending-torsion coupled beam and then extends it to model a non-uniform wing.

The governing partial differential equations of motion of the bending-torsion coupled beam (wing) shown in figure 1 are given by [9, 10]

$$EIh'''' + m\ddot{h} - mx_\alpha\ddot{\psi} = 0 \tag{1}$$
$$GJ\psi'' + mx_\alpha\ddot{h} - I_\alpha\ddot{\psi} = 0 \tag{2}$$

where *EI* and *GJ* are the bending and torsional rigidities of the beam, *m* is the mass per unit length, $I_\alpha$is the polar mass moment of inertia per length about the *Y*-axis and the primes and over dots denotes partial differentiation with respect to position *y* and time *t*, respectively.

For harmonic oscillation, sinusoidal variation in h and ψ with circular frequency ω may be assumed to give

$$h(y,t) = H(y)sin\omega t, \qquad \psi(y,t) = \Psi(y)sin\omega t \tag{3}$$

where $H(y)$ and $\Psi(y)$ denote the amplitude of the bending displacement and torsional rotation

Substituting equation (3) into equations (1) and (2) eliminates the time component and gives the following ordinary differential equations

$$EIH'''' - m\omega^2 H + mx_\alpha\omega^2\Psi = 0 \tag{4}$$
$$GJ\Psi'' + I_\alpha\omega^2\Psi - \omega^2 mx_\alpha H = 0 \tag{5}$$

where prime now denotes full differentiation with respect to *y*.

Equations (4) and (5) can be combined into a sixth order ordinary differential equation by eliminating either H or $\Psi$to give

$$W'''''' + \left(\frac{I_\alpha\omega^2}{GJ}\right)W'''' - \left(\frac{m\omega^2}{EI}\right)W'' - \left(\frac{m\omega^2}{EI}\right)\left(\frac{I_\alpha\omega^2}{GJ}\right)\left(\frac{I_\alpha - mx_\alpha^2}{I_\alpha}\right)W = 0 \tag{6}$$

where

$$W = H \; or \; \Psi \tag{7}$$

Equation (6) can be non-dimensionalised by using the non-dimensionalised length ξ where

$$\xi = \frac{y}{L} \tag{8}$$

Thus, with the help of equation (8), the non-dimensional form of equation (6) becomes

$$(D^6 + aD^4 - bD^2 - abc)W = 0 \tag{9}$$

where *a*, *b* and *c* are non-dimensional parameters given by

$$a = \left(\frac{I_\alpha \omega^2 L^2}{GJ}\right), \qquad b = \left(\frac{m\omega^2 L^4}{EI}\right), \qquad c = \left(\frac{I_\alpha - m x_\alpha^2}{I_\alpha}\right) \tag{10}$$

and D is the following differential operator

$$D = \frac{d}{d\xi} \tag{11}$$

The differential equation given by equation (9) can be solved using standard procedure [9, 10] to give

$$W(\xi) = C_1 cosh\alpha\xi + C_2 sinh\alpha\xi + C_3 cos\beta\xi + C_4 sin\beta\xi + C_5 cos\gamma\xi + C_6 sin\gamma\xi \tag{12}$$

where

$$\left.\begin{aligned} \alpha &= \left[2\left(\frac{q}{3}\right)^{\frac{1}{2}} cos\left(\frac{\phi}{3}\right) - \frac{a}{3}\right]^{\frac{1}{2}} \\ \beta &= \left[2\left(\frac{q}{3}\right)^{\frac{1}{2}} cos\left(\frac{(\pi-\phi)}{3}\right) + \frac{a}{3}\right]^{\frac{1}{2}} \\ \gamma &= \left[2\left(\frac{q}{3}\right)^{\frac{1}{2}} cos\left(\frac{(\pi+\phi)}{3}\right) + \frac{a}{3}\right]^{\frac{1}{2}} \end{aligned}\right\} \tag{13}$$

with

$$q = b + \frac{a^2}{3} \tag{14}$$

and

$$\phi = cos^{-1}\left[\frac{27abc - 9ab - 2a^3}{\left\{2(a^2+3b)^{\frac{3}{2}}\right\}}\right] \tag{15}$$

In equation (12), C_1-C_6 are the integration constants resulting from the solution of the governing differential equation (9).

$W(\xi)$ of equation (12) is the solution for both the bending displacement H and the torsional rotation Ψ, but with different sets of constants. Therefore,

$$H(\xi) = A_1 cosh\alpha\xi + A_2 sinh\alpha\xi + A_3 cos\beta\xi + A_4 sin\beta\xi + A_5 cos\gamma\zeta + A_6 sin\gamma\xi \tag{16}$$

and

$$\Psi(\xi) = B_1 cosh\alpha\xi + B_2 sinh\alpha\xi + B_3 cos\beta\xi + B_4 sin\beta\xi + B_5 cos\gamma\zeta + B_6 sin\gamma\xi \tag{17}$$

The two different sets of constants $A_1 - A_6$ and $B_1 - B_6$ in equations (16) and (17) can be related with the help of either equation (4) or equation (5) to give.

$$\left.\begin{aligned} B_1 &= k_\alpha A_1, \qquad B_3 = k_\beta A_3, \qquad B_5 = k_\gamma A_5 \\ B_2 &= k_\alpha A_2, \qquad B_4 = k_\beta A_4 \qquad B_6 = k_\gamma A_6 \end{aligned}\right\} \tag{18}$$

where

$$k_\alpha = \frac{b-\alpha^4}{bx_\alpha}, \quad k_\beta = \frac{b-\beta^4}{bx_\alpha}, \quad k_\gamma = \frac{b-\gamma^4}{bx_\alpha} \tag{19}$$

The expressions for bending rotation $\theta(\xi)$, bending moment $M(\xi)$, shear force $S(\xi)$ and torque $T(\xi)$ are given by

$$\theta(\xi) = H'\left(\frac{\xi}{L}\right) = \left(\frac{1}{L}\right)\{A_1 \alpha sinh\alpha\xi + A_2 \alpha cosh\alpha\xi - A_3 \beta sin\beta\xi + A_4 \beta cos\beta\xi - A_5 \gamma sin\gamma\xi + A_6 \gamma cos\gamma\xi\} \quad (20)$$

$$M(\xi) = -\left(\frac{EI}{L^2}\right) H''(\xi) = -\left(\frac{EI}{L^2}\right)\{A_1 \alpha^2 cosh\alpha\xi + A_2 \alpha^2 sinh\alpha\xi - A_3 \beta^2 cos\beta\xi - A_4 \beta^2 sin\beta\xi - A_5 \gamma^2 cos\gamma\xi - A_6 \gamma^2 sin\gamma\xi\} \quad (21)$$

$$S(\xi) = \left(\frac{EI}{L^3}\right)\{A_1 \alpha^3 sinh\alpha\xi + A_2 \alpha^3 cosh\alpha\xi + A_3 \beta^3 sin\beta\xi - A_4 \beta^3 cos\beta\xi + A_5 \gamma^3 sin\gamma\xi - A_6 \gamma^3 cos\gamma\xi\} \quad (22)$$

$$T(\xi) = \left(\frac{GJ}{L}\right) \Psi'(\xi) = \left(\frac{GJ}{L}\right)\{B_1 \alpha sinh\alpha\xi + B_2 \alpha cosh\alpha\xi - B_3 \beta sin\beta\xi + B_4 \beta cos\beta\xi - B_5 \gamma sin\gamma\xi + B_6 \gamma cos\gamma\xi\} \quad (23)$$

With the help of equations (16)-(23), the dynamic stiffness matrix of the coupled bending-torsion beam element which is essentially an aircraft wing element can be developed by applying the boundary conditions for displacements and forces at the ends of the elements.

Referring to figure 2, the boundary conditions for displacements are

$$\left.\begin{aligned} &\text{At } y = 0\ (\xi = 0):\ H = H_1, \quad \theta = \theta_1 \quad \Psi = \Psi_1 \\ &\text{At } y = L\ (\xi = 1):\ H = H_2, \quad \theta = \theta_2 \quad \Psi = \Psi_2 \end{aligned}\right\} \quad (24)$$

Similarly, referring to figure 3, the boundary conditions for the forces are

$$\left.\begin{aligned} &\text{At } y = 0\ (\xi = 0):\ S = S_1, \quad M = M_1 \quad T = -T_1 \\ &\text{At } y = L\ (\xi = 1):\ S = -S_2, M = -M_2 \quad T = T_2 \end{aligned}\right\} \quad (25)$$

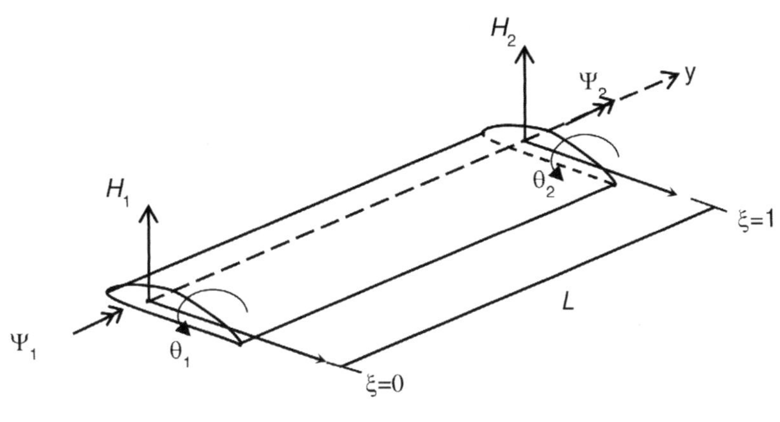

Figure 2. Boundary conditions for displacements of an aircraft wing element.

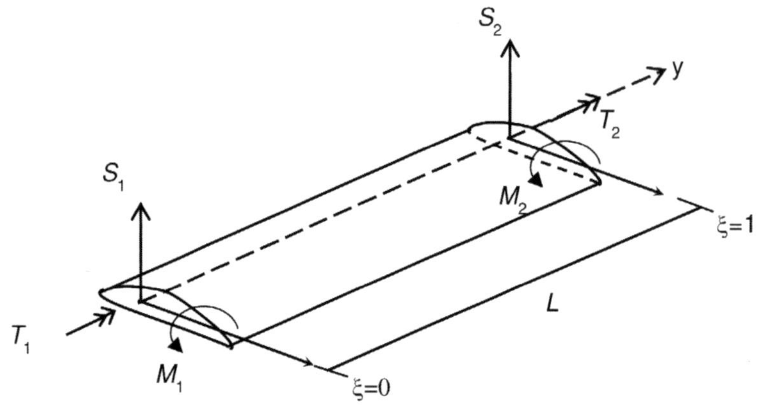

Figure 3. Boundary conditions for forces of an aircraft wing element.

Substituting the boundary conditions for displacements given by equation (24) into equations (16), (20) and (17), one obtains the following matrix relationship

$$\begin{bmatrix} H_1 \\ \theta_1 \\ \Psi_1 \\ H_2 \\ \theta_2 \\ \Psi_2 \end{bmatrix} = \begin{bmatrix} 1 & 0 & 1 & 0 & 1 & 0 \\ 0 & \alpha/L & 0 & \beta/L & 0 & \gamma/L \\ k_\alpha & 0 & k_\beta & 0 & k_\gamma & 0 \\ C_{h_\alpha} & S_{h_\alpha} & C_\beta & S_\beta & C_\gamma & S_\gamma \\ \alpha S_{h_\alpha}/L & \alpha C_{h_\alpha}/L & -\beta S_\beta/L & \beta C_\beta/L & -\gamma S_\gamma/L & \gamma C_\gamma/L \\ k_\alpha C_{h_\alpha} & k_\alpha S_{h_\alpha} & k_\beta C_\beta & k_\beta S_\beta & k_\gamma C_\gamma & k_\gamma S_\gamma \end{bmatrix} \begin{bmatrix} A_1 \\ A_2 \\ A_3 \\ A_4 \\ A_5 \\ A_6 \end{bmatrix} \tag{26}$$

or

$$\Delta = \mathbf{BA} \tag{27}$$

where **A** is the contact vector comprising the constants A_1 - A_6 and

$$C_{h\alpha} = \cosh\alpha;\ S_{h\alpha} = \sinh\alpha;\ C_\beta = \cos\beta;\ S_\beta = \sin\beta;\ C_\gamma = \cos\gamma;\ S_\gamma = \sin\gamma \tag{28}$$

Substituting the boundary conditions for forces given by equation (25) into equations (22), (21) and (23), one obtains the following matrix relationship

$$\begin{bmatrix} S_1 \\ M_1 \\ T_1 \\ S_2 \\ M_2 \\ T_2 \end{bmatrix} = \begin{bmatrix} 0 & W_3\alpha^3 & 0 & -W_3\beta^3 & 0 & -W_3\gamma^3 \\ -W_2\alpha^2 & 0 & W_2\beta^2 & 0 & W_2\gamma^2 & 0 \\ 0 & -W_1k_\alpha\alpha & 0 & -W_1k_\beta\beta & 0 & -W_1k_\gamma\gamma \\ -W_3\alpha^3S_{h_\alpha} & -W_3\alpha^3C_{h_\alpha} & -W_3\beta^3S_\beta & W_3\beta^3C_\beta & -W_3\gamma^3S_\gamma & W_3\gamma^3C_\gamma \\ W_2\alpha^2C_{h_\alpha} & W_2\alpha^2S_{h_\alpha} & -W_2\beta^2C_\beta & -W_2\beta^2S_\beta & -W_2\gamma^2C_\gamma & -W_2\gamma^2S_\gamma \\ W_1k_\alpha\alpha S_{h_\alpha} & W_1k_\alpha\alpha C_{h_\alpha} & -W_1k_\beta\beta S_\beta & W_1k_\beta\beta C_\beta & -W_1k_\gamma\gamma S_\gamma & W_1k_\gamma\gamma C_\gamma \end{bmatrix} \begin{bmatrix} A_1 \\ A_2 \\ A_3 \\ A_4 \\ A_5 \\ A_6 \end{bmatrix} \tag{29}$$

or

$$\mathbf{F} = \mathbf{D}\mathbf{A} \tag{30}$$

where

$$W_1 = \frac{GJ}{L};\ W_2 = \frac{EI}{L^2};\ W_3 = \frac{EI}{L^3} \tag{31}$$

The constant vector **A** can now be eliminated from equations (27) and (30) to give the following force-displacement relationship

$$\mathbf{F} = \mathbf{K}\,\Delta \tag{32}$$

where **K** is the 6×6 frequency dependent dynamic stiffness matrix given by

$$\mathbf{K} = \mathbf{D}\,\mathbf{B}^{-1} \tag{33}$$

The dynamic stiffness matrix of equation (33) representing a bending-torsion coupled beam such as an aircraft wing can now be used to model an aircraft wing. A non-uniform aircraft wing can be modelled as an assembly of many uniform dynamic stiffness elements. For instance, the unswept cantilever wing of figure 4 can be modelled as a stepped cantilever beam (wing) as shown in figure 5 where the non-uniform wing is split into 10 uniform dynamic stiffness elements. The dynamic stiffness elements of each of the 10 elements can be assembled to form the overall dynamic stiffness matrix of the complete wing. The straight unsweep wing and its idealisation in figures 4 and 5 are shown only for convenience, but the theory given above is sufficiently general and can handle swept and other wings with complex geometries.

The solution procedure to extract the natural frequencies and mode shapes from the overall dynamic stiffness matrix of the wing is based on the application of the Wittrick-Williams algorithm [8] which has featured in hundreds of papers. The algorithm is particularly suitable in solving free vibration problem using the dynamic stiffness method. The working principle of the algorithm is briefly summarised in the next section.

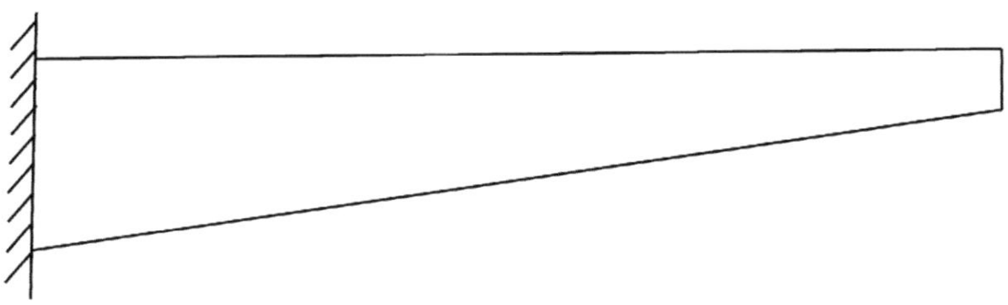

Figure 4. A non-uniform cantilever wing.

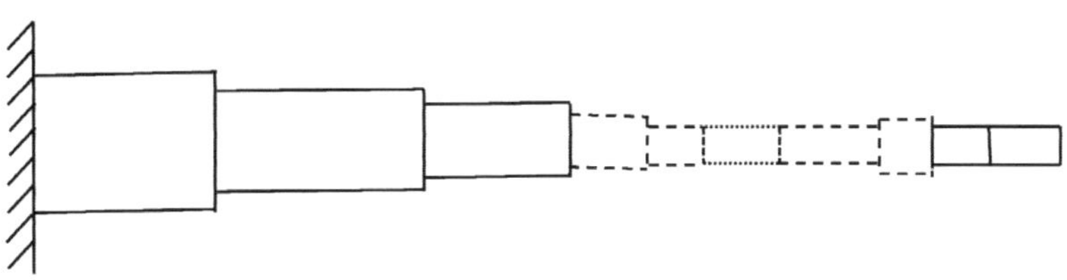

Figure 5. A non-uniform cantilever wing idealised as a stepped beam.

2.2 *Application of the Wittrick-Willams algorithm*

The dynamic stiffness matrix of equation (33) can now be used to compute the natural frequencies and mode shapes of aircraft wings. A non-uniform and/or swept wing can be analysed for its natural frequencies and mode shapes by idealising it as an assemblage of many uniform dynamic stiffness elements of bending-torsion coupled beams. The natural frequency calculation is accomplished by applying the Wittrick-Williams algorithm [8] which has received extensive coverage in the literature. Before applying the algorithm the dynamic stiffness matrices of all individual elements (see figures 4 and 5) need to be assembled to form the overall dynamic stiffness matrix $\mathbf{K}_f$ of the complete wing. The algorithm monitors the Sturm sequence condition of $\mathbf{K}_f$ in such a way that there is no possibility of missing any natural frequency of the wing. The application procedure of the algorithm is briefly summarised as follows.

Suppose that ω denotes the circular (or angular) frequency of the vibrating wing. Then according to the Wittrick-Williams algorithm [8], j, the number of natural frequencies passed, as ω is increased from zero to ω^*, is given by

$$j = j_0 + s\{\mathbf{K}_f\} \tag{34}$$

where $\mathbf{K}_f$, the overall dynamic stiffness matrix of the wing whose elements depend on ω is evaluated at $\omega = \omega^*$; $s\{\mathbf{K}_f\}$ is the number of negative elements on the leading diagonal of $\mathbf{K}_f^\Delta$, $\mathbf{K}_f^\Delta$ is the upper triangular matrix obtained by applying the usual form of Gauss elimination to $\mathbf{K}_f$, and j_0 is the number of natural frequencies of the wing still lying between $\omega = 0$ and $\omega = \omega^*$ when the displacement components to which $\mathbf{K}_f$ corresponds are all zeros. (Note that the structure can still have natural frequencies when all its nodes are clamped, because exact member equations allow each individual member to displace between nodes with an infinite number of degrees of freedom, and hence infinite number of natural frequencies between nodes.) Thus

$$j_0 = \sum j_m \tag{35}$$

where j_m is the number of natural frequencies between $\omega = 0$ and $\omega = \omega^*$ for an individual component member with its ends fully clamped, while the summation extends over all members of the structure. Thus, with the knowledge of equations (34) and (35), it is possible to ascertain how many natural frequencies of the wing lie below an arbitrarily chosen trial frequency (ω^*). This simple feature of the algorithm can be used to converge upon any required natural frequency to any desired accuracy. As successive trial frequencies can be chosen, computer implementation of the algorithm is very simple. However, for a detailed understanding, readers are referred to the original work of Wittrick and Williams [8].

3. Results and discussion

Using the above theory, two categories of aircraft wings with cantilever boundary condition at the root are analysed for their modal characteristics. In the first category, a class of high aspect ratio, high performance sailplane wings are considered. A typical layout of such a sailplane is shown in figure 6. Results for natural frequencies and mode shapes are computed for two sailplanes (S1 and S2) with spans 22m and 15m, respectively. Some particulars of the two sailplanes are given in Table 1. The second category of aircraft wings analysed belongs to transport airliners. A typical layout is shown in figure 7. Four wings of transport airliners (T1, T2, T3 and T4) with particulars given in Table 2 are analysed. In all cases, 10 dynamic stiffness elements were used to represent each wing. The data used for the stiffness (EI and GJ) and mass/inertia (m and I_α) properties of the wings and the shear centre locations (x_α) were calculated from the cross-sectional drawings of the wings expending considerable efforts. These data for the six aircraft are far too extensive to report in this paper.

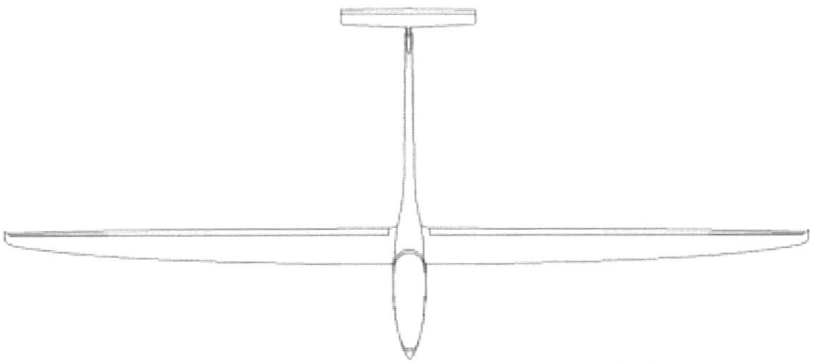

Figure 6. A general lay-out of a typical sailplane.

Figure 7. A general lay-out of a typical transport aircraft.

Table 1. Particulars of sailplanes

Parameters	Sailplane	
	Sailplane-S1	Sailplane-S2
Wing Span (m)	22	15
Wing Area (m^2)	15.44	10.05
Aspect Ratio	31.35	22.4
Wing Root Chord (m)	1.0	0.9
Wing Tip Chord (m)	0.4	0.4
Sweep angle (deg)	0	0
Length overall (m)	7.6	6.72
Height Overall (m)	2.0	2.0
Weight Empty (kg)	390	234
Max Take-off weight (kg)	550	440
Max Wing Loading (kg/m^2)	37	36
Max Cruising Speed (knots)	135	105

Table 2. Particulars of transport airliners.

Parameters	Transport airliner			
	T1	T2	T3	T4
Wing Span (m)	40	30	35	60
Wing Area (m^2)	162	93	123	362
Aspect Ratio	10	9	10	10
Wing Root Chord (m)	5.0	5.5	6.0	10.5
Wing Tip Chord (m)	2.5	1.5	1.5	2.5
Sweep angle (deg)	0	28	28	28
Length overall (m)	30	36	38	60
Height Overall (m)	12	11	12	17
Weight Empty (kg)	34,000	26,000	42,000	130,000
Max Take-off weight (kg)	70,000	46,000	74,000	275,000
Max Wing Loading (kg/m^2)	434	511	600	760
Max Cruising Speed (knots)	348	529	516	569
Range (nmi)	2835	2400	2592	8000

The first five natural frequencies of the six aircraft wings (two for sailplanes and four for transport airliners) are shown in Table 3. The letters B and T shown in the parenthesis indicate bending and torsion dominated modes, respectively whereas the letter C indicates a coupled mode with substantial amount of both bending and torsional displacements. It should be noted that sailplane wings do not carry engines whereas the transport airline wings have engine(s) with mass and inertia properties which have significant effects on natural frequencies. (Engine mass is a huge proportion of the total wing mass.) The mode shapes for the two sailplanes corresponding to the natural frequencies of Table 3 are shown in figure 8 whereas those of the four transport airliner wings are shown in figures 9 and 10, respectively.

Table 3. Natural frequencies of sailplane and transport airliner wings. (B): Bending dominated mode; (T): Torsional dominated mode;(C): Bending-Torsion coupled mode.

Aircraft Category	Natural Frequencies (ω_i) (rad/s)				
	ω_1	ω_2	ω_3	ω_4	ω_5
Sailplane S1	10.64(B)	42.62(B)	109.6(B)	111.5(T)	201.4(B)
Sailplane S2	13.38(B)	42.09(B)	93.35(B)	164.2(T)	167.4(C)
Transport Airliner T1	11.52(B)	33.09(B)	45.43(C)	87.85(B)	97.76(C)
Transport Airliner T2	19.71(B)	55.29(B)	100.2(B)	120.9(C)	197.7(C)
Transport Airliner T3	11.99(B)	34.69(B)	67.66(B)	74.14(T)	118.4(C)
Transport Airliner T4	8.994(B)	26.45(B)	47.64(T)	72.70(B)	94.64(T)

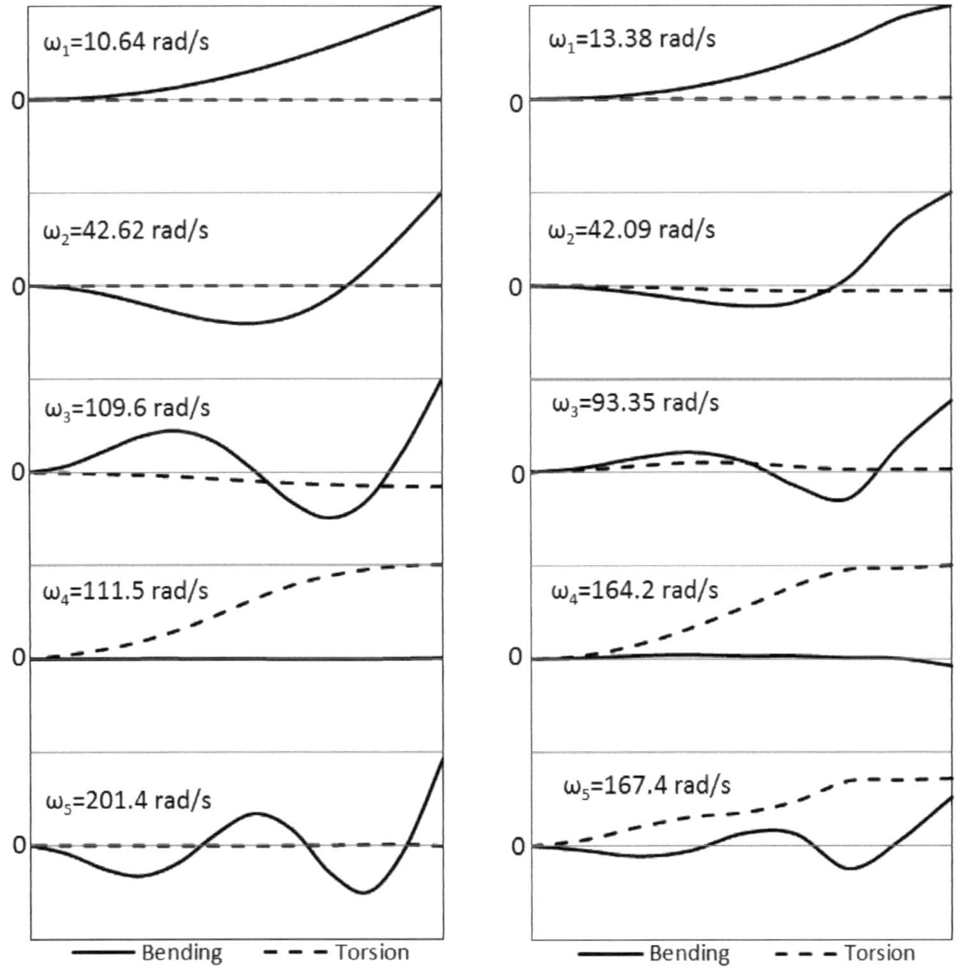

Figure 8. Natural frequencies and mode shapes of Sailplane wings S1 and S2.

Clearly, the first three modes of the cantilever wings of the two sailplanes S1 and S2 are bending modes whereas the fourth mode for each of them is a pure torsional mode, see figure 8. The fifth mode for the S1 wing is a bending mode. By contrast, for the S2 wing it is a coupled mode.

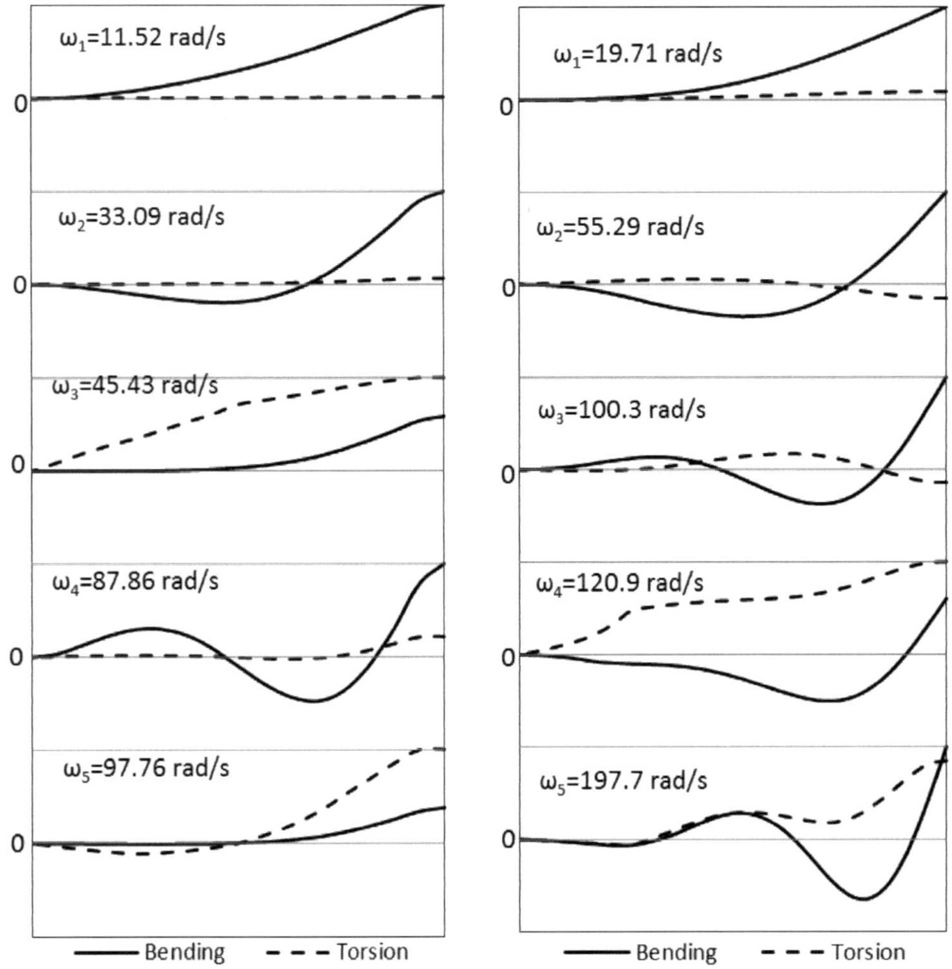

Figure 9. Natural frequencies and mode shapes of transport airliner wings T1 and T2.

Referring to Table 3 and figure 9, the first two modes of the cantilever wings of transport airliner T1 and T2 are essentially bending modes, but the nature of the third mode for the two wings differs quite significantly. For T1, it is basically a coupled mode dominated by torsional displacement, but for T2, it is a bending dominated mode. The fourth mode for T1 is bending dominated, but for T2 it is actually a bending-torsion coupled mode. To all intents and purposes, the fifth mode for both T1 and T2 is a coupled mode. Now referring to figure 10, the mode shapes for T3 and T4 wings are discussed. The first two modes for these two cantilever wings are essentially bending modes as was the case with T1 and T2 wings. However, the third mode for T3 and T4 are different. For the T3 wing, it is a bending-torsion coupled mode, but dominated by bending whereas for the T4 wing, it is a pure torsional mode. The fourth mode for the T3 wing is mainly a torsion dominated mode with some bending deformation present, but for the T4 wing it is a bending dominated mode with a small amount of torsion present. The fifth mode for T3 is a coupled mode whereas for T4, it is a torsional mode.

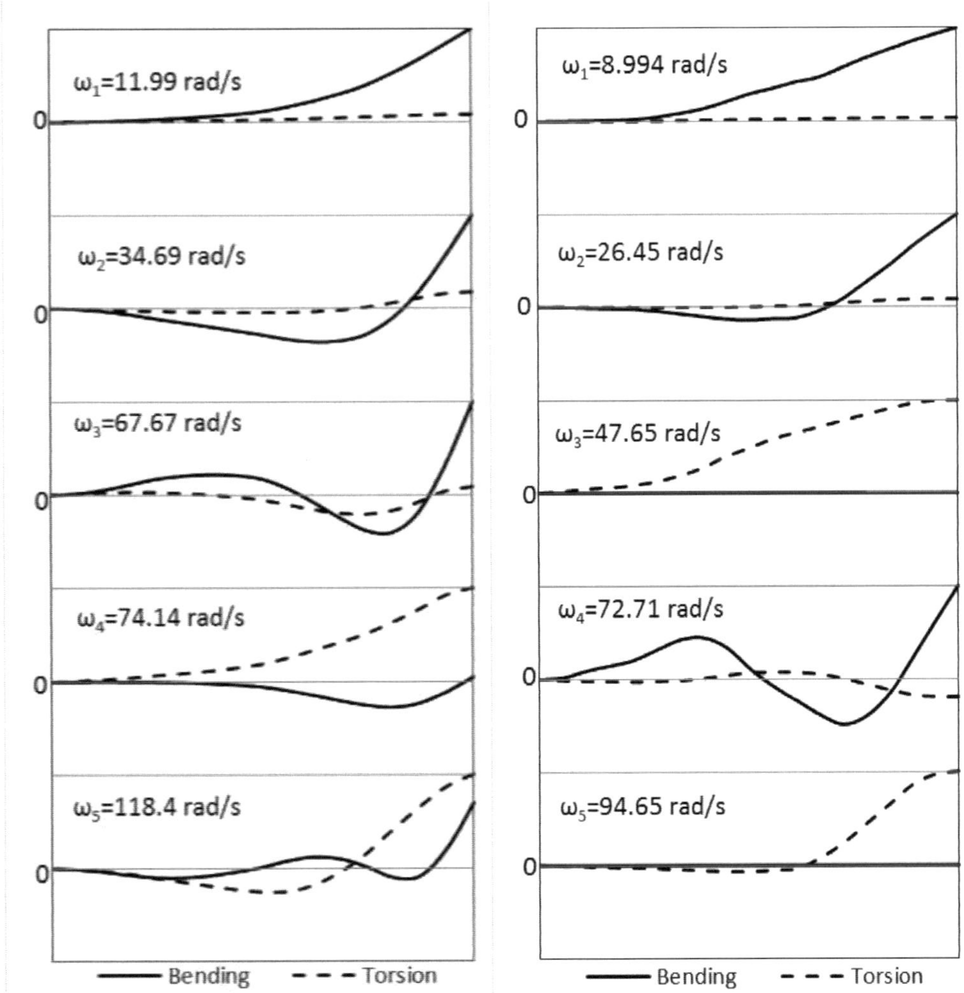

Figure 10. Natural frequencies and mode shapes of transport airliner wings T3 and T4.

4. Conclusions

Using the dynamic stiffness method and by applying the Wittrick-Williams algorithm, the modal behaviour of two sailplane and four transport aircraft wings is investigated. Natural frequencies and mode shapes for these wide ranging aircraft wings for cantilever boundary conditions are illustrated. The results are examined and discussed. In general, the first two modes for each of the six aircraft are effectively bending modes, but the third mode is either bending or torsional or a coupled mode depending on the type of the wing analysed. The fourth mode is again either torsion or bending dominated or even coupled. The same observation is made for the fifth mode. The investigation paves the way to establish trends for the modal behaviour of high aspect ratio aircraft wings.

Acknowledgements

The author is grateful to his PhD students Ajandan Ananthapuvirajah and Hassan Kassem for help given in the preparation of tables and graphs in this paper.

5. References

[1] Banerjee J R 1984 Flutter characteristics of high aspect ratio tailless aircraft *J. Aircraft* **21** 733-735.

[2] Banerjee J R 1988 Flutter modes of high aspect ratio tailless aircraft *J. Aircraft* **25** 473-476.

[3] van Schoor M C and von Flotow A H 1990 Aeroelastic characteristics of highly flexible aircraft, *J Aircraft* **27** 901-908.

[4] Eslimy-Isfahany S H R, Banerjee J R and Sobey A J 1996 Response of a bending-torsion coupled beam to deterministic and random loads *J. Sound Vib.* **195** 267-283.

[5] Banerjee J R, Patel M H, Done G T S, Butler R and Lillico M 1998 Free vibration and flutter sensitivity analyses of a large transport aircraft *Proc. 7th AIAA/USAF/NASA/ISSMO Sympos. on Multidisc. Analys. and Optimis,* St. Louis, Missouri, USA, Paper No. 98-4765.

[6] Tang D and Dowell E H 2001 Experimental and theoretical study on aeroelastic response of high aspect ratio wings, *AIAA* **39** 1430-1441.

[7] Banerjee J R, Liu X and Kassem H I 2014 Aeroelastic stability of high aspect ratio aircraft wings *J. Appl. Nonlin. Dyn.* **3** 413-422.

[8] Wittrick W H and Williams F W 1971 A general algorithm for computing natural frequencies of elastic structures *Quart. J. Mech. and Appl. Math.* **24** 263-284.

[9] Banerjee J R 1989 Coupled bending-torsional dynamic stiffness matrix for beam elements *Int. J. Num. Meth. Eng.* **28** 1283-1298.

[10] Banerjee J R 1991 A FORTRAN program for computation of coupled bending-torsional dynamic stiffness matrix of beam elements *Adv. Eng. Software*, **13** 17-24.

Self-Weight Loading of Horizontal Hydraulic Cylinders with Axial Load

J M Hale, Ee Yu Sim

School of Mechanical and Systems Engineering
Newcastle University
Newcastle upon Tyne
NE1 7RU
United Kingdom

E-mail jack.hale@ncl.ac.uk

Abstract. An iterative matrix method is described to determine the deformed shape of a horizontal stepped tube with large axial load, representing a heavy hydraulic cylinder. The method is applicable to both simply supported (pinned-pinned) and clamped-clamped end constraints. The clamped-clamped case is extended to include rotational compliance in the clamps. Using this analysis, radial loads on the piston seals of the cylinder are determined.

Comparison is made between the seal loading of a typical large cylinder in pinned-pinned and clamped-clamped configurations, respectively. It is shown that the seal loading can be considerably reduced by the use of the clamped-clamped configuration provided the ends can be well aligned, but that the load is sensitive to angular misalignment.

1. Introduction

Long telescopic hydraulic rams are restricted in their use by the large bending moment that occurs near the mid span as they start to buckle under the combined influence of self weight and axial load. This causes large radial loads on the seals, which in turn leads to rapid wear.

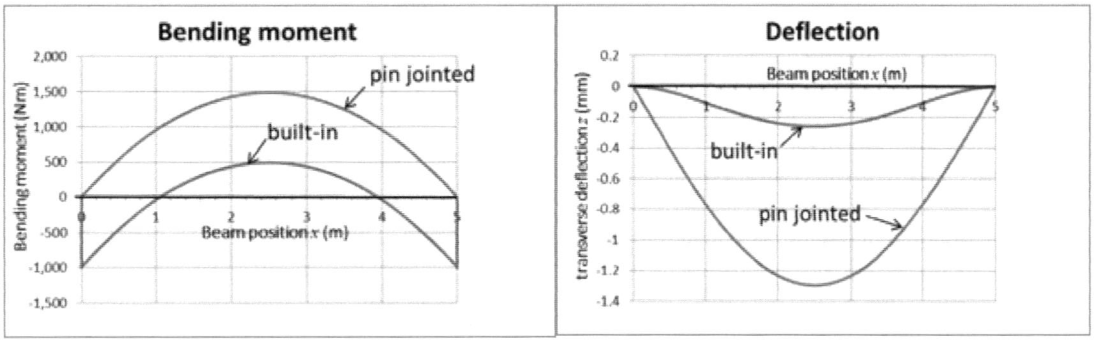

Figure 1. Deflection and bending moment of horizontal beams with self-weight loading

Mid-span deflection and bending moment are both greatly reduced by the use of clamped-clamped (built-in) end constraints.

Putting aside the effects of axial load and changing cross-section initially, the large bending moment in a horizontal beam due to self weight is of the form shown in Figure 1 for simply supported and clamped-clamped or built-in ends, respectively. This is reflected in the small normal deflection in the clamped-clamped case, also shown in Figure 1.

A method for determining bending moment and normal displacement of a stepped shaft under combined self weight and axial force with different end mounts is presented. This is a good idealisation of a hydraulic ram, and the seal loads are obtained from the bending moment at the step. Results are presented for a single stage ram, but the method is readily extended to multi-stage telescopic rams. The method used, based on iteration and numerical integration, accommodates any number of changes of section and/or material, and any combination of misalignment of the ends, and so goes beyond the closed form solution for a uniform bar without misalignment used for validation. Furthermore, it lends itself to extension to the dynamic problem of a long horizontal ram subject to imposed vertical motion, for instance in a heavy earthmoving vehicle.

2. Analysis of a stepped beam with self-weight loading

A single stage ram loaded by self-weight only may be idealised as a stepped beam as shown in Figure 2 for the clamped-clamped case. It may be analysed using the conventional double integration method [1] by making imaginary cuts and applying a shear force F and bending moment M to maintain equilibrium as shown in Figure 3. M_A and R_A are the reactions at the clamped left hand end. w_1 and w_2 are the self weights per unit length of the two sections.

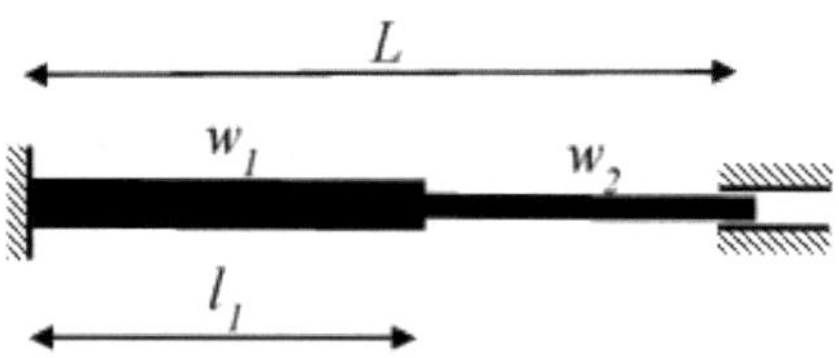

Figure 2. Stepped beam with self-weight

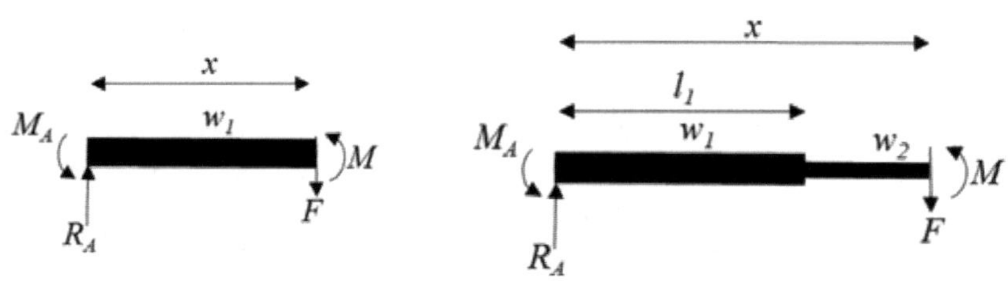

(a) left hand section (b) right hand section

Figure 3. Loading on a two section beam with self weight

Taking moments about the imaginary cut face, an expression may be obtained for bending moment as a function of the applied loads including the end constraints. The bending moment is related to the curvature by:

$$M = EI\frac{d^2z}{dx^2}$$

where M is the local bending moment
E is young's modulus
I is second moment of area
x is distance from the end of the beam
z is normal deflection of the beam (a function of x

Journal of Physics: Conference Series **721** (2016) 012006 doi:10.1088/1742-6596/721/1/012006

For the left hand region, Figure 3(a):

$$M + M_A - R_A x + \frac{1}{2}w_1 x^2 = 0 \quad (1)$$

$$\therefore \frac{d^2z}{dx^2} = \frac{1}{E_1 I_1}M = \frac{1}{E_1 I_1}\left(-M_A + R_A x - \frac{1}{2}w_1 x^2\right) \quad (2)$$

Integrating: $$\frac{dz}{dx} = \frac{1}{E_1 I_1}\left(-M_A x + \frac{1}{2}R_A x^2 - \frac{1}{6}w_1 x^3 + C_1\right) \quad (3)$$

and $$z = \frac{1}{E_1 I_1}\left(-\frac{1}{2}M_A x^2 + \frac{1}{6}R_A x^3 - \frac{1}{24}w_1 x^4 + C_1 x + C_2\right) \quad (4)$$

Similarly, for the right hand section in Figure 3(b):

$$M + M_A - R_A x + w_1 l_1\left(x - \frac{1}{2}l_1\right) + \frac{1}{2}w_2 l_1 (x - l_1)^2 = 0 \quad (5)$$

$$\therefore \frac{d^2z}{dx^2} = \frac{1}{E_2 I_2}M = \frac{1}{EI}\left(-M_A + R_A x - w_1 l_1\left(x - \frac{1}{2}l_1\right) - \frac{1}{2}w_2 (x - l_1)^2\right) \quad (6)$$

$$\therefore \frac{dz}{dx} = \frac{1}{E_2 I_2}\left(-M_A x + \frac{1}{2}R_A x^2 - \frac{1}{2}w_1 l_1\left(x - \frac{1}{2}l_1\right)^2 - \frac{1}{6}w_2 (x - l_1)^3 + D_1\right) \quad (7)$$

$$\therefore z = \frac{1}{E_2 I_2}\left(-\frac{1}{2}M_A x^2 + \frac{1}{6}R_A x^3 - \frac{1}{6}w_1 l_1\left(x - \frac{1}{2}l_1\right)^3 - \frac{1}{24}w_2 (x - l_1)^4 + D_1 x + D_2\right) \quad (8)$$

In equations 1-8 E, I and w take the suffix 1 or 2 to indicate values for the 1st (left hand) and 2nd (right hand) sections of the ram.

There are four constants of integration and two unknown reactions M_A and R_A. These may be evaluated from six simultaneous equations: four boundary conditions z and $\frac{dz}{dx}$ at $x = 0$ and $x = L$, and equating displacement z and slope $\frac{dz}{dx}$ at $x = l_1$ in the two sections.

The values of the end fixing constraints M_A and R_A and the constants of integration can thus be obtained for any mounting configuration. If required, the constraints at the right hand end can then be obtained from equilibrium.

It is convenient to set up these simultaneous equations in matrix form:

$$\left(\begin{bmatrix} 0 & 0 & 1 & 0 & 0 & 0 \\ 0 & 0 & 0 & 1 & 0 & 0 \\ 0 & 0 & 0 & 0 & 0 & 0 \\ 0 & 0 & 0 & 0 & 0 & 0 \\ -l_1 & \frac{1}{2}l_1^2 & 1 & 0 & 0 & 0 \\ -\frac{1}{2}l_1^2 & -\frac{1}{6}l_1^3 & l_1 & 1 & 0 & 0 \end{bmatrix}\begin{pmatrix} M_A \\ R_A \\ C_1 \\ C_2 \\ D_1 \\ D_2 \end{pmatrix} + \begin{Bmatrix} 0 \\ 0 \\ 0 \\ 0 \\ -\frac{1}{6}w_1 l_1^3 \\ -\frac{1}{24}w_1 l_1^4 \end{Bmatrix}\right)\frac{1}{E_1 I_1}$$

$$= \left(\begin{bmatrix} 0 & 0 & 0 & 0 & 0 & 0 \\ 0 & 0 & 0 & 0 & 0 & 0 \\ -L & \frac{1}{2}L^2 & 0 & 0 & 1 & 0 \\ -\frac{1}{2}L^2 & \frac{1}{6}L^3 & 0 & 0 & L & 1 \\ -l_1 & \frac{1}{2}l_1^2 & 0 & 0 & 1 & 0 \\ -\frac{1}{2}l_1^2 & -\frac{1}{6}l_1^3 & 0 & 0 & l_1 & 1 \end{bmatrix}\begin{pmatrix} M_A \\ R_A \\ C_1 \\ C_2 \\ D_1 \\ D_2 \end{pmatrix} + \begin{Bmatrix} 0 \\ 0 \\ -\frac{1}{2}w_1 l_1\left(L - \frac{1}{2}l_1\right)^2 - \frac{1}{6}w_2 (L - l_1)^3 \\ -\frac{1}{6}w_1 l_1\left(L - \frac{1}{2}l_1\right)^3 - \frac{1}{24}w_2 (L - l_1)^4 \\ -\frac{1}{8}w_1 l_1^3 \\ -\frac{1}{48}w_1 l_1^4 \end{Bmatrix}\right)\frac{1}{E_2 I_2}$$

The left hand side is the left hand region (equations 3-4) and the right hand side is the right hand region (equations 7-8).

Rows 1 and 2 are the slope and displacement at the left hand end $x = 0$ (equations 3 and 4).

Rows 3 and 4 are the slope and displacement at the right hand end $x = L$ (equations 7 and 8).

Rows 5 and 6 are the slope and displacement at the discontinuity $x = l_1$ (equations 3 and 7).

This matrix equation may be written as $[K_1]\{V\} + \{K_2\} = [K_5]\{V\} + \{K_6\}$ (9)
Rearranging:

$$([K_1] - [K_5])\{V\} = \{K_6\} - \{K_2\}$$

$[K_1] - [K_4]$ is a square matrix and is readily inverted, and $\{K_6\} - \{K_2\}$ is a column vector, so the equation (9) becomes:

$$\{V\} = ([K_1] - [K_5])^{-1}(\{K_6\} - \{K_2\}) \tag{10}$$

and solved for M_A, R_A and the constants of integration C_1, C_2 D_1, D_2 using MatLab [2].

For any more complex stepped beam, the same analysis may be used. Two new constants of integration are introduced for each step, but there are also two new equilibrium equations, and so the system may be solved in the same way.

It should be noted that this problem of the stepped beam cannot be analysed by Macaulay's method. This makes use of equating slope and displacement at loading discontinuities and uses a mathematical trick to make the constants of integration the same in all regions. This breaks down for a stepped bar because the modulus of rigidity *EI* is different in each region.

For comparison, a similar method was used to analyse a simply supported stepped beam. For this case the slope and deflection at $x = l_1$ are equated for the two regions as before and so rows 5 and 6 of the matrix equation are unchanged. The boundary condition $z = 0$ at the right hand and left hand ends ($x = 0$ and $x = L$) and so rows 2 and 4 are also unchanged. The slopes at the ends are unknown, so rows 1 and 3 are replaced by the bending moment equations 2 and 6 with *M* set to zero at $x = 0$ and $x = L$.

This method was validated by making the stepped beam appear as a uniform beam and comparing the results with standard solutions. This was done in three ways:

1. the values of *E*, *I* and *w* were made the same in regions 1 and 2;
2. the length of the left hand section l_1 was set to zero;
3. the length of the left hand section l_1 was set to *L*.

In each case the bending moments at the ends and mid-span, and the deflection at mid-span, agreed with standard solutions for the equivalent uniform beam.

3. Analysis of a heavy hydraulic ram with end loading

Adding axial load F_A as shown in Figure 4 changes the loading on an imaginary cut face as shown in Figure 5.

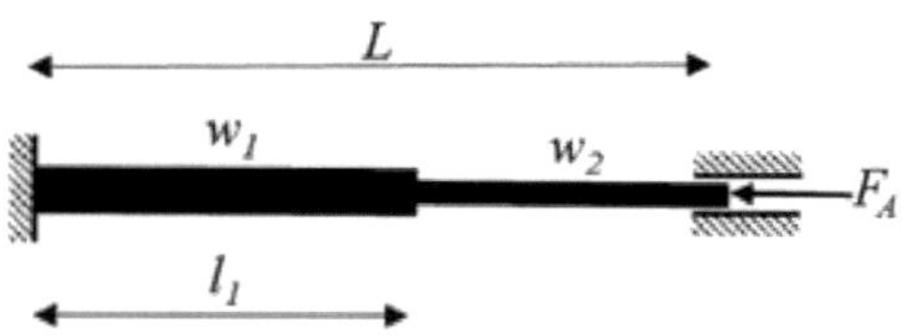

Figure 4. Loading on a two section clamped-clamped beam with self weight and axial force F_A

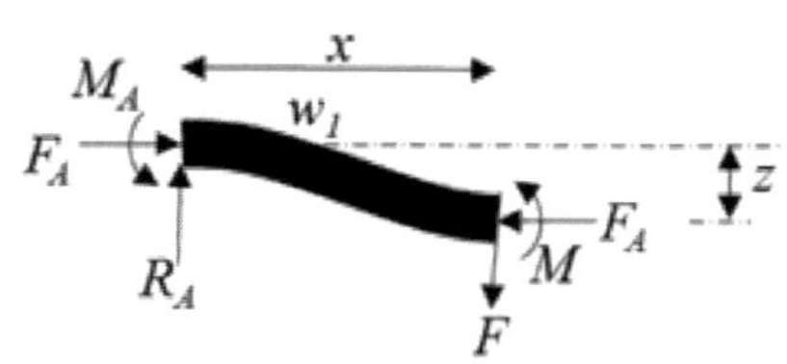

Figure 5. Equilibrium of a section of the stepped beam with self weight and axial force F_A

This has the effect of adding an extra term $+F_A z(x)$ to equations (1) and (5). This means that the integrations cannot be done directly because the displacement $z(x)$ is a function of those integrations. The best we can say is that the slope equations (3) and (7) take an extra term $+F_A \int_0^x zdx$ and the displacement equations (4) and (8) an extra term $+F_A \iint_0^x zdxdx$.

An iterative method has been found to solve these equations effectively. Starting from a "reasonable" deformed shape $z(x)$, the integral terms $F_A \int_0^x zdx$ and $F_A \iint_0^x zdxdx$ were evaluated by numerical integration (trapezium rule). These integral values were added to the matrix equation (9) by adding to appropriate rows in vectors $\{K_3\}$ and $\{K_7\}$, forming the equation:

$$[K_1]\{V\} + \{K_2\} + \{K_3\} = [K_5]\{V\} + \{K_6\} + \{K_7\} \tag{11}$$

It was found that the equations converged provided the end load was less than the buckling load. The method was validated by degenerating the stepped beam into a uniform beam by the three methods described above and comparing the results for bending moment and deflection at the centre span with the results given in Roark's "Formulas for Stress and Strain" table 10.2e [3].

Results are presented for a 5m long single stage hydraulic ram. The theory above was embodied in a MatLab program which calculates the deformation and bending moment, and hence radial load on the seals at the change of section. The ram has dimensions:

cylinder O/D 200mm,
piston O/D 188mm,
wall thickness of cylinder and piston 6mm.

The cylinder and piston material is steel of density 7860kg/m^3 and Young's modulus 210GPa, filled with oil of density 910kg/m^3. The axial force is 10^6N (1.0MN), which is approaching the buckling load for the pin jointed cylinder.

The overlap between piston and cylinder at full extension is taken as 200mm with a seal at each end. From this, the radial load on the seals is easily calculated from the bending moment at the change of section.

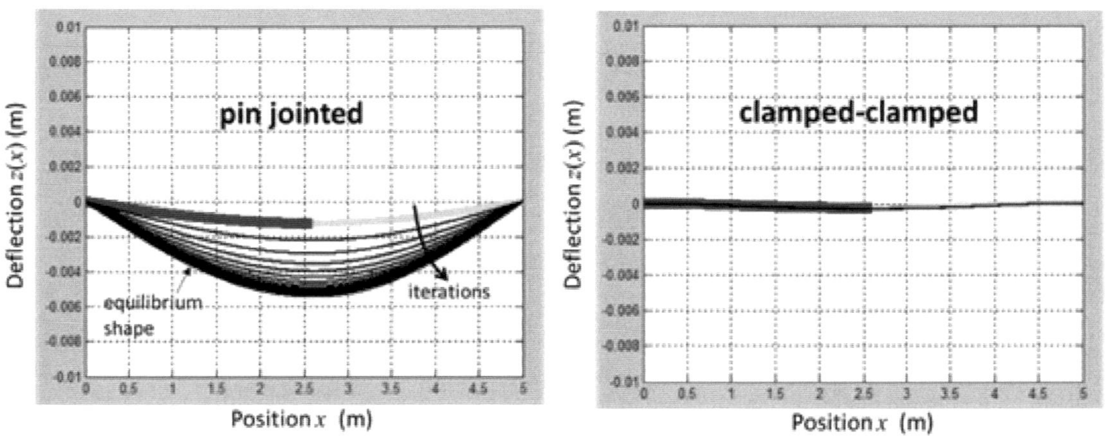

Figure 6. Displacement of stepped beam with self-weight and axial loading

Figure 6 shows the effect of this large axial load of 1MN, close to the buckling load for the pin jointed beam. It may be seen that it has a large effect on the pin jointed beam, increasing the maximum deflection from 1.3mm to 5.4mm. The mid-span bending moment is also greatly increased, raising the radial load on the seal from 3.9kN to 35kN. However, the axial load has very little effect on the clamped-clamped beam, increasing the maximum deflection by only 0.1mm from 0.2 to 0.3mm. The seal load is increased from 2.6 to 3.4kN.

It is evident that the use of the clamped-clamped arrangement has a big effect, reducing both mid-span deflection and seal loading by an order of magnitude.

4. The effects of misalignment

Pin jointed cylinders are unaffected by misalignment of the ends because they are self-aligning. However, clamped ends force the cylinder to take up a shape determined by the slope and position of the ends, and this is reflected by the bending moment distribution, and hence the radial loads on the seals.

To investigate this effect, equations 3, 4, 7 and 8 (slope and deflection at $x = 0$ and $x = L$) must be modified. Equations 3 and 4 refer to slope and deflection at the left hand end, so the expressions are equated to imposed values rather than zero. According to the convention that the calculated values for the left hand section of the ram (section 1) go on the left hand side of the matrix equation, the imposed values are added to rows 1 and 2 of the right hand side.

Similarly, in equations 7 and 8 the imposed slope and deflection (θ_B and z_B) at the far right of the ram ($x = L$) are added to rows 3 and 4 on the left hand side of the matrix equation. Equating slope and deflection at the change of section ($x = l_1$) is unaffected by misalignment and so rows 5 and 6 of the matrix equation are unchanged.

For convenience, these misalignment values are placed in new column vectors $\{K_4\}$ and $\{K_8\}$ to form an enlarged matrix equation:

$$[K_1]\{V\} + \{K_2\} + \{K_3\} + \{K_4\} = [K_5]\{V\} + \{K_6\} + \{K_7\} + \{K_8\} \quad (12)$$

$$\text{where } \{K_4\} = \begin{Bmatrix} 0 \\ 0 \\ \theta_B \\ z_B - z_A \\ 0 \\ 0 \end{Bmatrix} \text{ and } \{K_8\} = \begin{Bmatrix} \theta_A \\ 0 \\ 0 \\ 0 \\ 0 \\ 0 \end{Bmatrix}$$

Note that for convenience, the origin of the co-ordinate system is taken as the position of the left hand end of the cylinder irrespective of its misalignment.

As before, equation 12 may be rearranged to:

$$([K_1] - [K_5])\{V\} = \{K_6\} + \{K_7\} + \{K_8\} - \{K_2\} - \{K_3\} - \{K_4\} \quad (13)$$

and solved for $\{V\}$ to obtain the constants of integration and the constraint moments and reactions, and hence the distribution of bending moment from which the loads on the seals are calculated.

The effect of an axial misalignment of 9mm is shown in Figure 7(a) for the same 5m cylinder with 1MN axial load. It may be seen that deformed shape is hardly affected by the axial load, which is well below the buckling load for the clamped beam. The calculated radial seal load is 3.8kN. As shown in Figure 7(b), radial seal load is proportional to axial misalignment, but for reasonable offsets below about 25mm there is very little effect, and the seal load only approaches that of the pinned-pinned case at offset misalignments greater than 800mm.

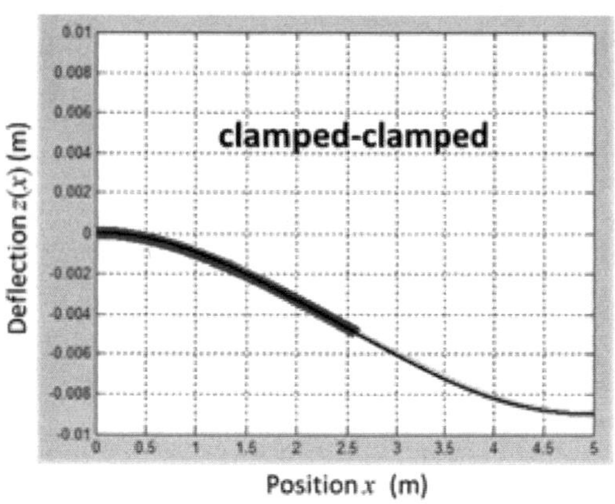

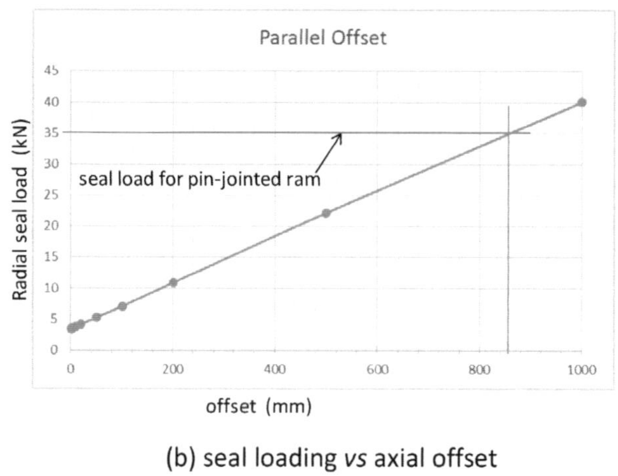

Figure 7. Stepped beam with axial (parallel) misalignment. Loading by self-weight and axial force.

The effect of angular misalignment of 0.4° at one end, shown in Figure 8(a), has more dramatic effect. Again the radial seal load is proportional to misalignment, but as shown in Figure 4(b), the seal load rises very quickly with increased misalignment angle and reaches the value for the pin jointed case at this misalignment of only 0.4°.

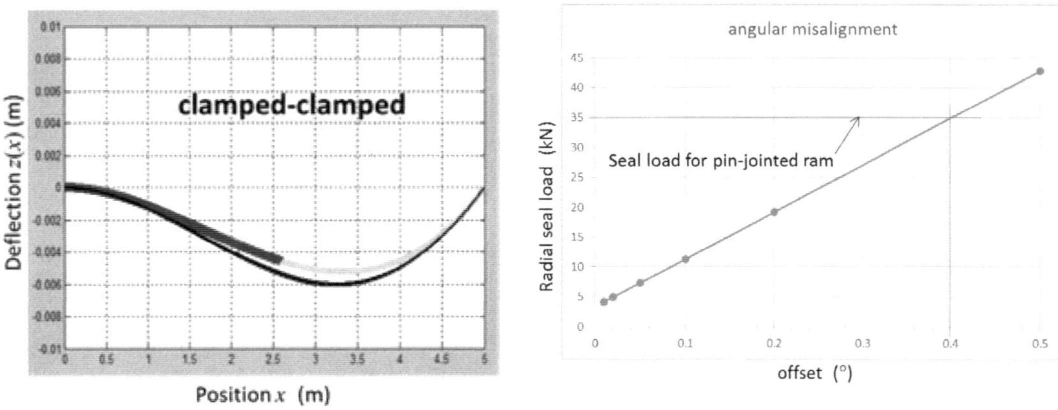

(a) displacement (b) seal loading *vs* angular misalignment

Figure 8. Stepped beam with angular misalignment.
Loading by self-weight and axial force.

5. Compliance

A further complication is the impossibility of providing truly "clamped" conditions, in which there is no displacement of rotation of the cylinder ends under load. From the analysis above, displacement parallel to the cylinder axis due to the end reactions will have negligible effect, but rotation due to the fixing moments is likely to have a significant effect on the seal loads. Estimates of this effect may be obtained from Roark's "Formulas for Stress and Strain" table 24.20b [3]. For the 200mm diameter cylinder used in the examples above, welded to a circular 8mm thick steel plate of 400mm diameter simply supported at its edge, the compliance would be 1.6×10^{-6} rad/Nm or 90×10^{-6} degrees/Nm.

For rotationally compliant end fixings the slope of the cylinder at the ends becomes a function of both angular misalignment and fixing moment. At the left hand end ($x = 0$), equation 1 becomes $M + M_A = 0$, and so the bending moment at the left hand end of the cylinder is $-M_A$. Angular compliance c is angle rotated per unit applied moment, so the slope of the cylinder due to compliance $\alpha_A = -cM_A$ and this must be added to the slope due to angular misalignment. This is achieved by adding a term $-c$ to location (1,1) in the matrix $[K_5]$ of equation 13.

Similarly, at the right hand end ($x = L$), equation 5 gives

$$M + M_A - R_AL + w_1l_1\left(L - \frac{1}{2}l_1\right) + \frac{1}{2}w_2l_1(L - l_1)^2 = 0$$

where the bending moment M becomes the fixing moment M_B.

$$\therefore M_B = -M_A + R_AL - w_1l_1\left(L - \frac{1}{2}l_1\right) - \frac{1}{2}w_2l_1(L - l_1)^2$$

The slope due to compliance is $\alpha_B = -cM_B$. This is incorporated into row 3 of the matrix equation by inserting c and $-cL$ to locations (3,1) and (3,2), respectively, of the matrix $[K_1]$ and inserting $c\left(w_1l_1\left(L - \frac{1}{2}l_1\right) + \frac{1}{2}w_2l_1(L - l_1)^2\right)$ into row 3 of vector $\{K_2\}$.

This formulation was tested within the MatLab program and validated by degenerating the stepped beam into a uniform beam as before. With the compliance c set to zero the beam behaved as a clamped-clamped beam and with it set to any large value greater than 0.5°/Nm it behaved as a simply supported beam.

The MatLab program was used to investigate the effect of moderate compliance on the behaviour of the misaligned 5m × 200mm diameter stepped cylinder with clamped-clamped ends. As might be expected, the compliance tends to influence the seal loading towards that in the simply supported case. For large misalignment, where the alignment angle is greater than the natural angle taken up in the simply supported case (0.4° for the case studied), this reduces the loading on the seal, but gives little improvement below 10^{-3}degree/Nm. For small misalignment less than the simply supported slope, the increase in seal load is modest (<7%) for compliance below 10^{-5}degree/Nm.

For this case of the 200mm diameter cylinder considered here, a compliance of 10^{-5}degree/Nm is very small (i.e. it is very stiff), being provided by a 400mm diameter 12mm thick steel plate clamped around its rim, or a 15mm thick plate simply supported. 10^{-3}degree/Nm is still quite small (i.e. quite stiff), being provided by a 650mm diameter 5mm thick steel plate with clamped support.

6. Discussion of results

It is clear from the results presented that a significant reduction in loading on the seals can be obtained by the use of clamped ends, providing the angular misalignment can be kept small and the clamping is stiff (low compliance).

If the compliance cannot be kept low, then clamping might still be a lower cost solution than pined joints and would still give a somewhat reduced seal loading.

If angular alignment cannot be maintained, then a pinned solution will always be required.

7. Conclusions

1. An iterative method has been established to analyse the internal loading and distortion of a stepped beam loaded with combined self weight and axial force.
2. It is demonstrated that a significant reduction in maximum bending moment is obtained by constraining the ends (clamped-clamped support).
3. This reduced bending moment is translated directly into reduced radial seal loading in hydraulic cylinders compared with the conventional pin jointed ends.
4. This seal load reduction is largely independent of offset misalignment.
5. The load reduction is highly sensitive to angular misalignment, increasing to that in the pin-pin cylinder when the misalignment angle equals the slope taken up in the pin-pin case, and exceeding it at higher misalignment angles.
6. Compliance in the clamped supports increases the seal loads at small misalignment angles, but small compliance coupled with small angular misalignment gives significantly reduced seal loading.

References

[1] Ryder G H 1969 *Strength of Materials* MacMillan
[2] Biran Adrian and Breiner Moshe 1995 *MATLAB for Engineers* Adison-Wesley
[3] Young Warren C 1989 *Roark's Formulas for Stress and Strain (6th edition)* McGraw-Hill

The effect of time delay on control stability of an electromagnetic active tuned mass damper for vibration control

A Hassan, A Torres-Perez, S Kaczmarczyk, P Picton

The University of Northampton, St. George's Avenue, Northampton, NN2 6JD, United Kingdom

Ali.hassan@northampton.ac.uk

Abstract. The aim of this paper is to investigate the effect of time delays on the stability of a zero-placement position and velocity feedback law for a vibratory system comprising harmonic excitation equipped with an electromagnetic active tuned mass damper (ATMD). The purpose of the active control is broadening the vibration attenuation envelope of a primary mass to a higher frequency region identified as from 50±0.5Hz with a passive tuned mass damper (TMD) to a wider range of 50±5Hz with an ATMD. Stability conditions of the closed-loop system are determined by studying the position of the system closed-loop poles after the introduction of time delays for different excitation frequencies. A computer simulation of the model predicted that the proposed control system is subject to instability after a critical time delay margin dependent upon the frequency of excitation and the finding were experimentally validated. Three solutions are derived and experimentally tested for minimising the effect of time delays on the stability of the control system. The first solution is associated with the introduction of more damping in the absorber system. The second incorporates using a time-delayed ATMD by tuning its original natural resonant frequency to beyond the nominal operational frequency range of the composite system. The third involves an online gain tuning of filter coefficients in a dual arrangement of low-pass and high-pass filters to eliminate the effect time delays by manipulating the signal phase shifts

1. Introduction

The motivation of this work arises from the vibration problem that arises within a free piston Stirling engine generator. Stirling engines are power machines that operate over a closed, regenerative thermodynamic cycle, with cyclic compression and expansion of the working fluid at different temperature levels [1]. The power piston inside the engine is fitted with a set of permanent magnets that oscillate inside the linear alternator to generate electrical current at 50Hz. Hence this type of grid connected machines is designed and tuned to operate within a frequency range of 50Hz ± 0.5Hz in Europe. The reciprocating movement of the power piston causes a vibration within the engine that can be intolerable. In general, the damping of vibration in equipment is usually accomplished by employing visco-elastic materials [2,3]. Currently, a passive TMD is employed to absorb the vibration coming from the piston at 50Hz. Recently the European Network of Transmission System Operators for Electricity (ENTSO) has brought new regulations for grid connection and disconnection of low power generation devices which introduced wider grid operation frequencies between 47-53Hz. The operation of the Stirling engine is affected by the new regulation since its passive TMD is originally tuned to cancel the vibration of the engine within a very narrow bandwidth. Active damping techniques can achieve far better performance than simple passive ones by employing actuators and complex control systems. The active control of vibration has gained some considerable level of attention in civil engineering and a number of successful implementation was achieved [4]. In [5], an active tuned mass damper that could be used to control lateral vibrations induced by pedestrians on a bridge deck was presented. The work was tested using a bridge model equipped with a prototype active mass damper. In recent years, there has been a growing interest in using electromagnetic actuators instead of piezoelectric actuators. In [6] a four-mount active vibration isolation system using voice coil actuators was developed. In [7] a voice coil actuator with velocity feedback control for highly sensitive instruments by producing a sky-hook damper at low frequencies (2-6 Hz) was considered. Their proposed method reduced the vibration at resonance without causing larger vibration amplitudes at higher frequencies.

Closed-loop control using structural displacement and velocity feedbacks for active vibration reduction is perceived as a common practice in vibration control. It is generally difficult to achieve accurate measurements of velocity and displacement directly in many systems particularly that lack inertial reference frames. Acceleration measurements, on the other hand, are practically easier to obtain without the requirement of a reference frame and with sensors featuring accurate and wide frequency band [8]. Numerical integration is often employed in order to obtain displacement and velocity measurement from acceleration signals, resulting in the appearance of DC offset component in the signal, particularly after the second integration action [9]. Anti-aliasing and noise filtering, as well as DC component elimination induce an inevitable group time delay that may affect in deterring the control performance and even causing instability [10].

In terms of active vibration control studies that bear attention to the effect of time delays, a method of active control based on time delay compensation for the low-order modes of a cantilever-like structure suffering from complex excitations was developed in [11]. The solution was based on constructing an autoregressive moving average (ARMA) model for eliminating the time delay caused by filters. A control system for a time-delayed active vibration absorber, which can eliminate the vibration of the primary structure subject to a sinusoidal disturbance was proposed and tested experimentally in [12]. Two kinds of controller design methods for active vibration control problem with input time delay, where the time-delayed system is first transformed into a delay-free system, so that the classical control strategies such as the LQR method can be applied directly was developed in [13]. In [14] a time-delayed acceleration feedback controller for the active vibration control of flexible structure was proposed. In [15] a study on the design of the acceleration-based active vibration control for flexible structures with bonded piezoelectric patches is presented, where by manipulating the phase shift, time delay compensation caused by non-collocated sensor/actuator pair is achieved.

The work done in [16] serves a basis for the study presented in this paper. The active vibration control of a structure using displacement and velocity feedback law to change the natural resonance and the damping in an ATMD is implemented using a test rig that utilizes a voice coil motor for an ATMD. A detailed theoretical study into the effect of time delays on the control stability is presented by studying the system poles with the introduction of delays. The theoretical findings and experimental realization illustrated that instability of the control system occurs beyond a critical time delay margin depending on the excitation frequency. Two solutions to mitigate the time-delay induced instability were proposed and experimentally verified. The first is based on using a time-delayed ATMD whilst shifting its original resonant frequency above the system's nominal frequency band of operation. The second solution is based on an online gain tuning of the coefficients of a dual filtering arrangement in order to manipulate the individual phase shifts of the two filters accordingly so that a time delay effect is eliminated. It was found that the first solution incurs the use of more electrical power than the second and the hence it was found to be 50% less efficient in terms of power consumption.

2. Test rig development

Following the results of the theoretical work that was carried out in [16], a scaled test rig was developed to demonstrate the concept of active damping with ATMD with particular emphasis on the effect of time delays on the proposed control strategy

2.1. Test rig system

In the test rig shown in **Figure 1**, an electrodynamic shaker (V406) from B&K is mounted vertically as shown. The shaker acts as the source of sinusoidal excitation to the 2-DOF system which features a shaker table whose vibration to be cancelled by means of an ATMD system that, in turn, is passively tuned at 50Hz. The ATMD consists of a planar spring attached to the top of the moving shaker table. A Voice Coil Motor (VCM) provides the actuation for active damping and is mounted between the shaker table and the dynamic mass of the ATMD. The vibration model of the test rig is also shown in **Figure 1** on the left. The detailed description of the model is presented in the following section.

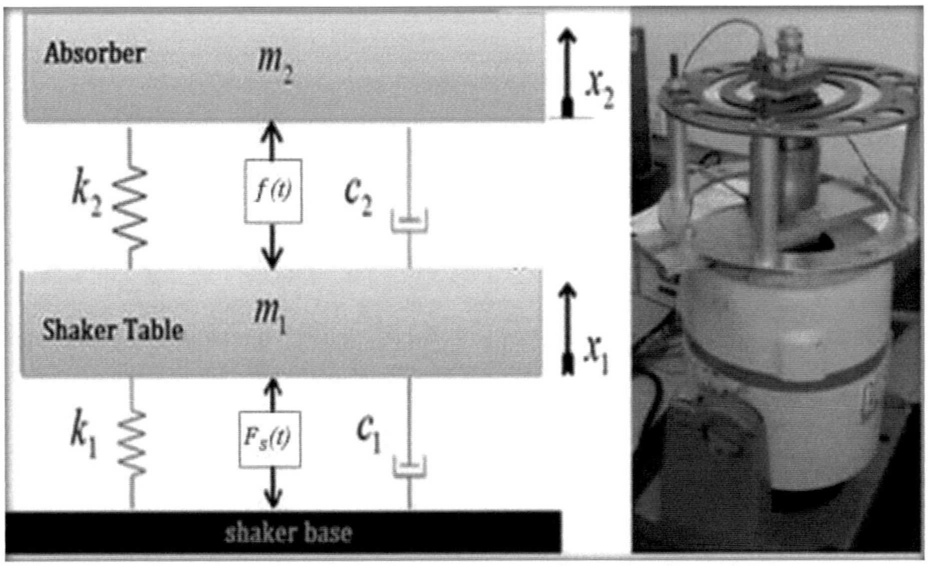

Figure 1. Test rig and the corresponding vibration model

2.2. Mathematical Model

An electromechanical model is developed to represent the shaker rig mathematically for the purpose of performing theoretical studies. The VCM is modelled as an RL circuit with V_a and i_a representing the voltage across and the current through the VCM terminals. The terms R_a, L_a, and k_a represent the motor coil inductance, resistance, and the motor constant respectively. The mass m_1 is the mass of the shaker table plus that of the VCM's stationary base and the absorber's stationary mass. The mass m_2 includes the dynamic mass of the absorber and the VCM's moving coil. Here it is assumed that the value of the back electromotive force (EMF) constant is the same as k_a. The control force f_a is proportional to the VCM motor current i_a by k_a. A proportional controller is used to control the VCM current i_a by comparing it to a desired current i_r . The excitation force F is generated by the dynamic shaker which has a similar model to the VCM represented with subscript s. The variable x_1 represents the shaker table displacement, x_2 the absorber mass displacement, $\dot{x}_1$ the shaker table velocity and $\dot{x}_2$ the absorber velocity. The resulting equations of the coupled electromechanical system are shown below

$$\begin{cases} m_1\ddot{x}_1 + k_1x_1 + c_1\dot{x}_1 + k_2(x_1 - x_2) + c_2(\dot{x}_1 - \dot{x}_2) = F_s - f_a \\ m_2\ddot{x}_2 + k_2(x_2 - x_1) + c_2(\dot{x}_2 - \dot{x}_1) = f_a \\ V_a = R_a i_a + L_a \dfrac{di_a}{dt} + k_a(\dot{x}_1 - \dot{x}_2) \\ V_s = R_s i_s + L_s \dfrac{di_s}{dt} - k_s\dot{x}_1 \\ F_s = k_s i_s \\ f_a = \beta(\dot{x}_1 - \dot{x}_2) + \gamma(x_1 - x_2) \\ i_r = f_a/k_a \\ V_a = k_p(i_r - i_a) \end{cases} \tag{1}$$

where γ and β are relative positon and relative velocity feedback gains for the zero placement control technique and derived as

$$\gamma = 4\pi^2\, m_2({f_{exc}}^2 - {f_0}^2) \qquad \text{and} \qquad \beta = -c_2 \tag{2}$$

f_{exc} is the excitation frequency and f_0 is the resonant frequency of the absorber. The stability condition of the control law is derived following the Routh-Hurwitz criterion [17]:

$$\alpha > -m_2 \qquad \beta > -c_2 \qquad \gamma > -k_2$$

For ATMD applications, maximum attenuation is achieved in the primary mass when the damping in the absorber is minimal; hence a negative velocity feedback gain equivalent to the damping coefficient of the absorber seems a perfect choice. However, by examining the stability condition of the control law, the velocity feedback gain β has to be greater than $-c_2$. Hence, to avoid a marginal stability situation, β is chosen slightly greater than $-c_2$ to achieve both, high attenuation level and stability. The following shows one of the possible state-space model representations of the 2-DOF system comprising the electrodynamic shaker and the ATMD.

$$\begin{bmatrix} \dot{x}_1 \\ \dot{x}_2 \\ \dot{x}_3 \\ \dot{x}_4 \\ \dot{x}_5 \\ \dot{x}_6 \\ \dot{x}_7 \\ \dot{x}_8 \\ \dot{x}_9 \end{bmatrix} = \begin{bmatrix} 0 & 0 & 0 & 0 & 1 & 0 & 0 & 0 & 0 \\ 0 & 0 & 0 & 0 & 0 & 1 & 0 & 0 & 0 \\ 0 & 0 & 0 & 0 & 0 & 0 & 1 & 0 & 0 \\ 0 & 0 & 0 & 0 & 0 & 0 & 0 & 1 & 0 \\ \frac{-(k_1+k_2)}{m_1} & \frac{k_2}{m_1} & 0 & 0 & \frac{-(c_1+c_2)}{m_1} & \frac{c_2}{m_1} & \frac{-k_a}{m_1} & \frac{k_s}{m_1} & 0 \\ \frac{k_2}{m_2} & \frac{-k_2}{m_2} & 0 & 0 & \frac{c_2}{m_2} & \frac{-c_2}{m_2} & \frac{k_a}{m_2} & 0 & 0 \\ 0 & 0 & 0 & 0 & \frac{k_a}{L_a} & \frac{-k_a}{L_a} & 0 & 0 & \frac{1}{L_a} \\ 0 & 0 & 0 & 0 & -\frac{k_s}{L_s} & 0 & 0 & \frac{-R_s}{L_s} & 0 \\ 0 & 0 & 0 & 0 & 0 & 0 & 0 & 0 & 0 \end{bmatrix} \begin{bmatrix} x_1 \\ x_2 \\ x_3 \\ x_4 \\ x_5 \\ x_6 \\ x_7 \\ x_8 \\ x_9 \end{bmatrix} + \begin{bmatrix} 0 & 0 \\ 0 & 0 \\ 0 & 0 \\ 0 & 0 \\ 0 & 0 \\ 0 & 0 \\ 0 & \frac{-R_a}{L_a} \\ \frac{1}{L_s} & 0 \\ 0 & 0 \end{bmatrix} \begin{bmatrix} V_s \\ i_a \end{bmatrix} \tag{3}$$

The following define the state-space variables

$x_1 = shaker\ table\ displacement$ $\quad x_2 = absorber\ displacement$ $\quad x_3 = electrical\ charge\ in\ the\ VCM$
$x_4 = electrical\ charge\ in\ the\ shaker\ coil$ $\quad x_5 = shaker\ table\ velocity$ $\quad x_6 = absorber\ velocity$
$x_7 = VCM\ current$ $\quad x_8 = shaker\ current$ $\quad x_9 = VCM\ voltage$

2.3. Model simulation and validation

In this section, the model of the test rig is simulated in the frequency domain and the results are plotted against the experimental results. The values of Table 1 are adopted for the model simulation and test parameters of the rig.

Table 1. Model Simulation parameters

Parameter	Value	Unit	Parameter	Value	Unit
m_1	2.1	[kg]	V_s	$3\sin(2\pi ft)$	[V]
m_2	0.403	[kg]	L_s	1.8	[mH]
c_1	1	[Ns/m]	R_s	1.2	[Ω]
c_2	7	[Ns/m]	k_s	11.1	[N/A]
k_1	12300	[N/m]	L_a	1.3	[mH]
k_2	39775	[N/m]	R_a	2.7	[Ω]
f_{exc}	5-100	[Hz]	k_a	8	[N/A]

In the experiment the VCM is not powered and the leads are left open- that is to say that there is no control operation throughout this test. The test is carried using a sinusoidal excitation with variable frequency of the shaker between 5Hz and 100Hz in open loop. Two accelerometers of 100 mv/g sensitivity and 70g max dynamic range were used to collect data from both the absorber and shaker table accelerations. In addition to that, the electrical current consumed by the shaker is measured using a low noise non-contact current of dynamic range $\pm 20A$ and 0.01 A resolution. The results from both the simulation and experiments are depicted in the graphs of Figure 2.

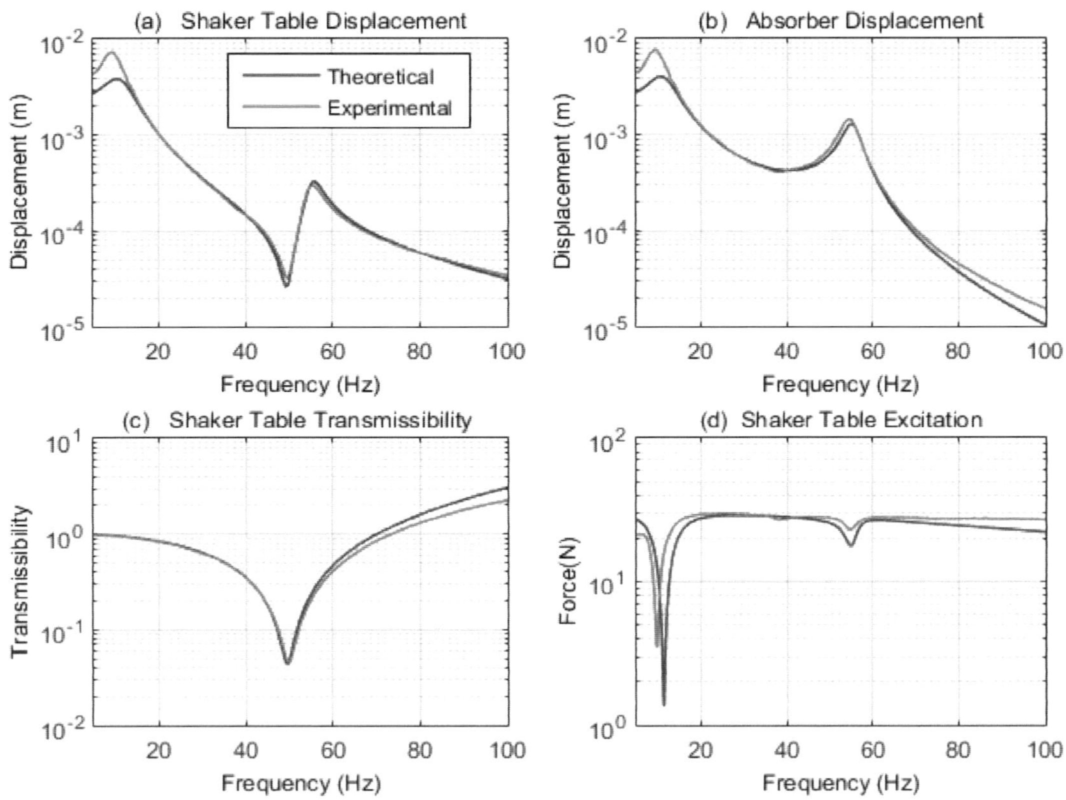

Figure 2. Theoretical model results vs experimental results. The value of damping coefficient is determined $c_2 = 7\ NS/m$

The results reveal a very close match between the theoretical model predictions and the experimental results. Figure 2 (a) shows the curves of the shaker table displacement with two peaks corresponding to the resonant frequencies of the composite at 10Hz and 55Hz. At 50Hz, the shaker displacement is attenuated to minimum due to the effect of the anti-resonant frequency of the TMD. The transmissibility to the shaker table is shown in (c) and reveals that at 50Hz, the shaker force is completely transmitted from the shaker table to the absorber mass. The experimental results revealed the VCM introduces damping into the ATMD system and the value of the damping is determined as $c_2 = 7\ Ns/m$ following graph reverse fitting of the theoretical results onto the experimental ones.

3. Time delays

3.1. Expression of the actuator force with time delays

The intention of this section is to represent actuator force with the time delay term embedded in it. This analysis considers the zero-placement control law with position and velocity feedback for the active control. A time delay $d(t-\tau)$ can be expressed by $e^{-\tau s}$ in the Laplace domain with τ defined as the amount of time delay. To make it possible for studying the poles of the transfer function, the exponential term of the delay has to be transformed into a rational form. The first order Taylor approximation assuming a sinusoidal force of period T in the frequency domain becomes

$$e^{-\tau s} \approx 1 - \tau s \tag{4}$$

The form of the delayed actuator force can be expressed as follows

$$f_a(t-\tau) = d(t-\tau)[\beta(\dot{x}_1 - \dot{x}_2) + \ \gamma(x_1 - x_2)] \tag{5}$$

Equation (5) is incorporated into the system model presented in system of equations (1) to account for delays in the system. While ignoring the shaker dynamics, the resulting transfer function from shaker excitation force to the shaker table displacement can expressed in the Laplace domain as follows

$$W(s) = \frac{X_1(s)}{F_s(s)} = \frac{b_3 s^3 + \ b_2 s^2 + \ b_1 s^1 + \ b_0}{a_4 s^4 + a_4 s^4 + a_3 s^3 + a_2 s^2 + a_1 s + a_0} \tag{6}$$

The coefficient of the numerator and the denominator of $W(s)$ are shown in *Appendix A*. The roots of $W(s)$ are studied and the real part of the closed-loop poles are investigated for different time delays as the excitation frequency is varied. The simulation was carried out for time delays between 0s and 200μs and excitation frequency between 45Hz and 55Hz. The values of **Table 1Error! Reference source not found.** are adopted for this simulation. The position gain γ and velocity gain β are calculated as per equation (2). Figure 3 below shows the variation of the real part of the closed-loop poles. The horizontal axis of each subplot represents that of the time delay in seconds and the vertical axis represents the real part of the poles. The imaginary part is ignored and only the real part of the system poles is considered. This analysis is similar to the conventional root-locus analysis but it doesn't consider the imaginary part. In fact it is not possible to perform this analysis with the conventional root-locus method because each value of time delay corresponds to a different transfer function $W(s)$.

The results presented below show that all the system poles have negative real parts for time delays between 0s and 200μs for excitation frequencies below 50Hz indicating to system stability. As the excitation frequency moves above the original resonance of the ATMD (50Hz), the control system becomes unstable shortly after a certain delay margin which decreases consistently with the increase in the excitation frequency.

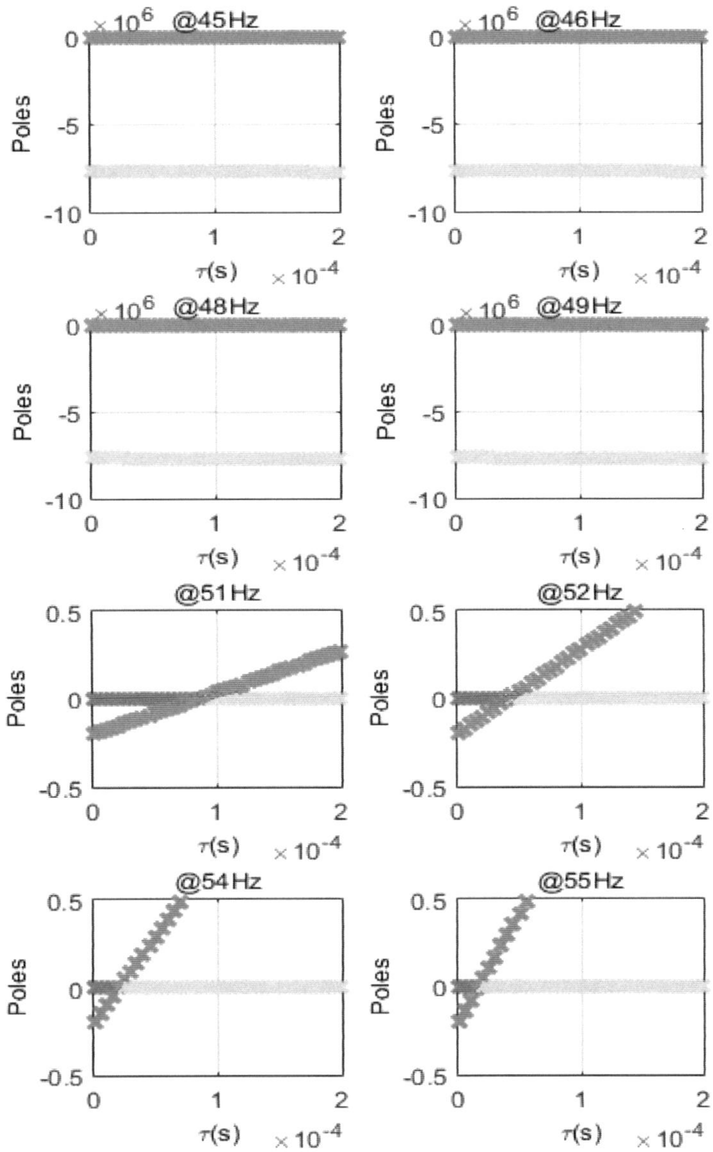

Figure 3. Plots showing the variation of the real part of the system poles as a function of time delay and excitation frequency with ATMD tuned at 50Hz with damping removed completely.

3.2. Solution 1: Change of damping in the ATMD

Another important observation is made concerning the relation between the critical time delay margin and the damping in the ATMD. The theoretical analysis shows that allowing a little more damping in the absorber improves the delay time margin as depicted in **Figure 4**. In this simulation, the velocity feedback gain beta was chosen to be $\beta > -c_2$ for the purpose of allowing more damping in the ATMD system. The results of this simulation are plotted in **Figure 4**. It is shown that when a little damping is allowed in the system, time delay still affect system stability however the time delay margin is longer at each particular frequency above 50Hz compared to **Figure 3**. While allowing more damping in the ATMD would eventually increases the stability margin, there exists an inverse trade-off between the amount of damping that is allowed in the ATMD and level of attenuation achieved in the vibration of the structure.

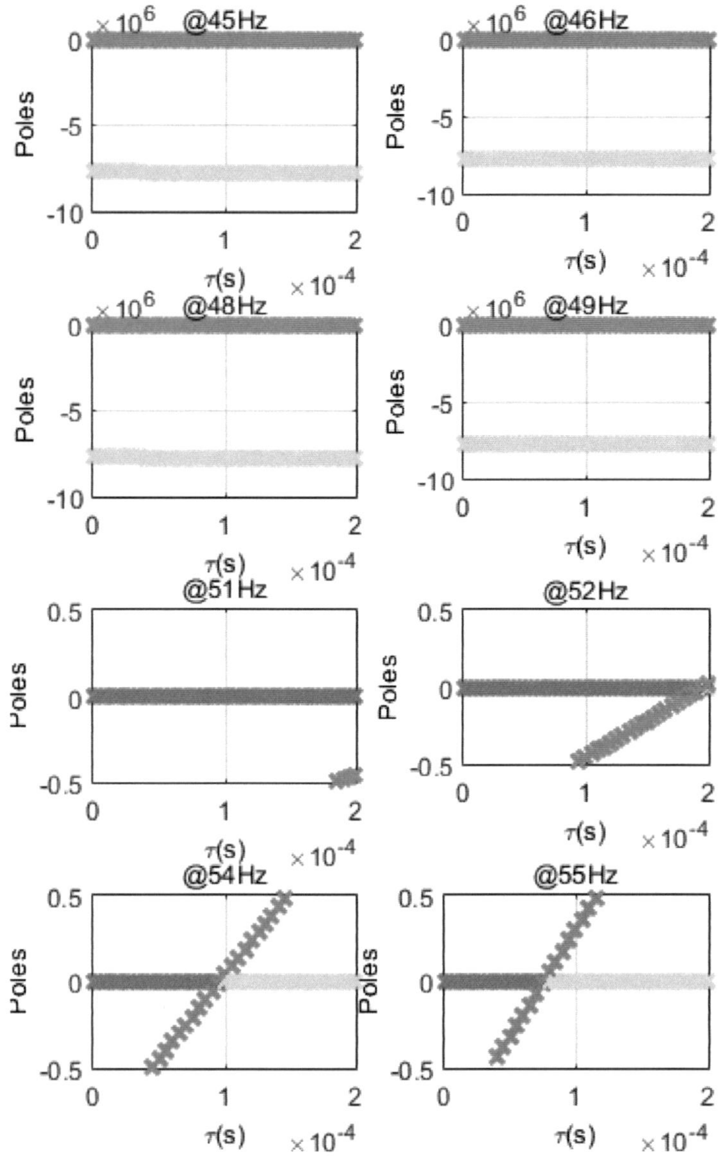

Figure 4 System poles variation as a function of time delay and excitation frequency with ATMD tuned at 50Hz with some damping allowed in the ATMD

3.3. Solution 2: Change of ATMD original resonance

By recalling that the nominal operational frequency range of the system is between 45Hz-55Hz, and by noticing that the system becomes unstable for time delay margins for excitation frequencies above the original tuning of the absorber, a solution can be formulated based on shifting the original tuning of the ATMD is shifted to a frequency higher than the original resonance 55Hz and outside the nominal frequency range of the engine operation, the system is guaranteed to be stable when the frequency of excitation is below the 55Hz. A new simulation is carried out to check the stability when the ATMD resonance is moved to 55Hz. The results are shown in Figure 5 and reveal that this technique succeeds in mitigating the effect of delays on stability.

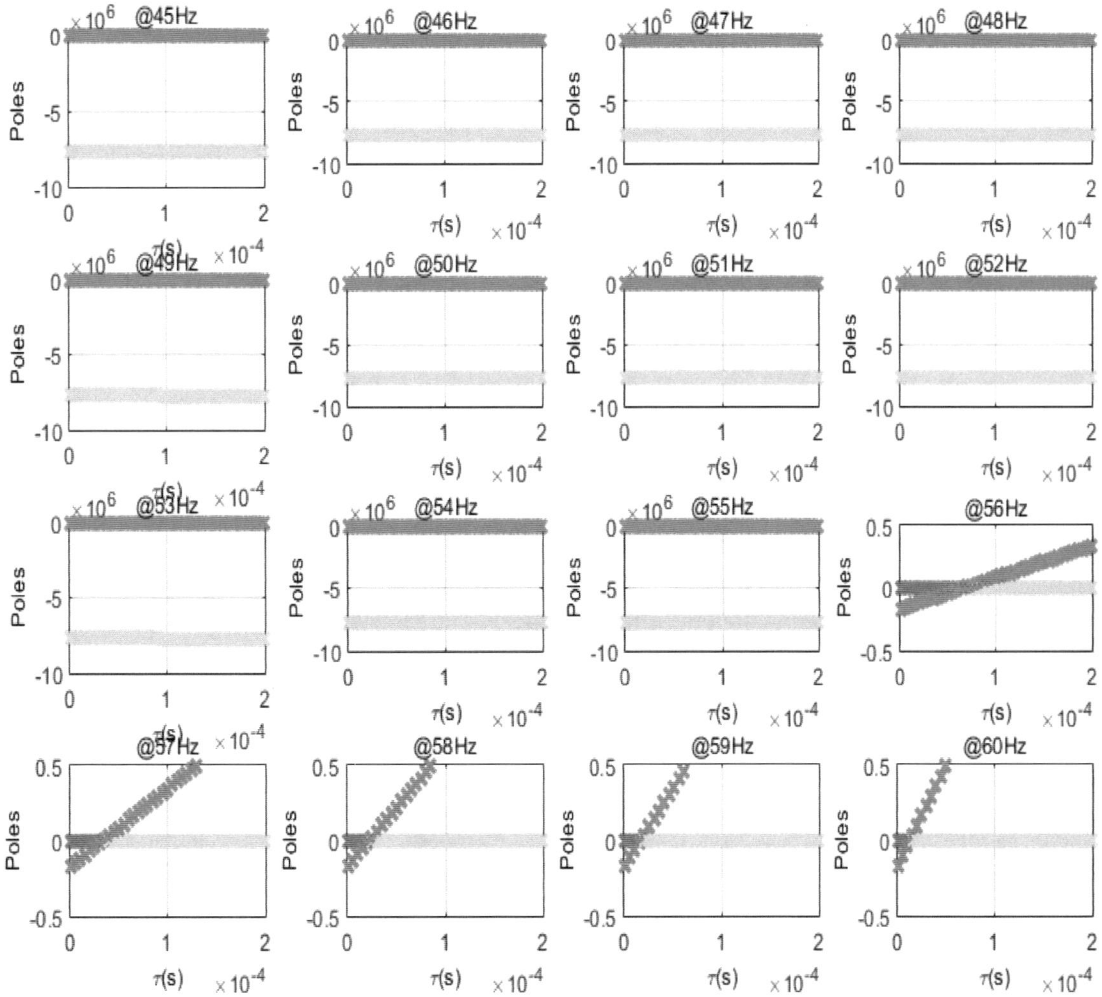

Figure 5. System poles variation as a function of time delay with 55Hz ATMD tuning

3.4. Discussion

While both of the techniques presented in the previous sub-sections serve as a successful mitigation solution of the time delay effect on stability of the control, there exist two main drawbacks that arise with following these approaches concerning efficiency and power consumption. The first technique works on the basis of increasing the time delay stability margin (time-delayed ATMD) by allowing for more damping in the system. Assuming the system has inherent damping such as the VCM, this technique is a feasible however a compromise on the level of attenuation of the shaker table vibration is made. The second technique also benefits from the design of a time-delayed ATMD, however damping can be removed from the system and hence maximum attenuation is possible. The main drawback with the second method arises from the perspective of power consumption since if the ATMD was originally tuned at 55Hz the position gain γ would be larger than the case when the ATMD was originally tuned at 50Hz as shown in and consequently more force is needed to actively shift the frequency of the ATMD.

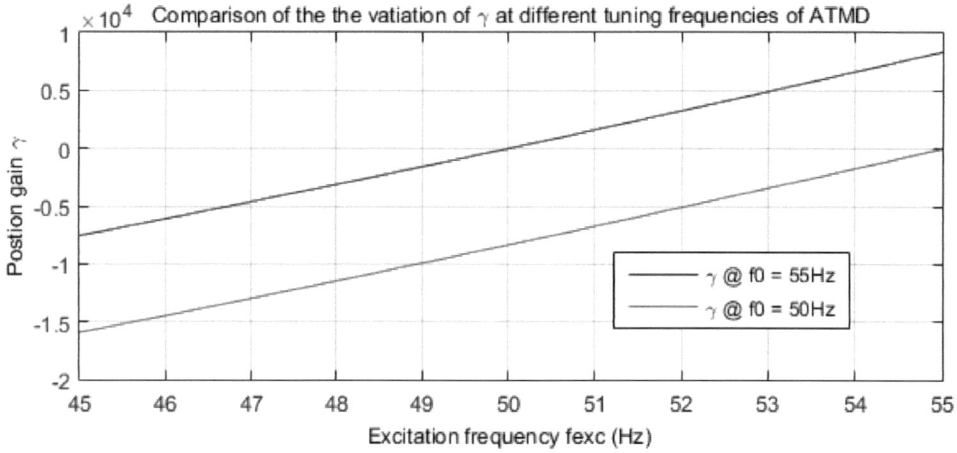

Figure 6. Position gain as a function of excitation frequency for two different ATMD resonant frequencies

3.5. Solution 3: Dual-filtering phase compensation

By recognizing the strengths and limitations of the previous two techniques, another method is proposed in this section for mitigating the effect of time delays on stability of the control system. The system under study is a linear-time invariant system with sinusoidal excitation and the steady-state of this system is assumed to be well known. In practice, the control implementation generally requires filtering of input signals for noise avoidance, anti-aliasing, and for DC component elimination. These induce inevitable group time delay that was shown earlier to affect stability of the control system. Furthermore, sensor measurement as well as non-collocated sensor/actuator pair in addition to processing times can induce more time delays. Reasons for these delays are geometrical arrangements, and computational time in analog digital converters (ADCs) and in digital analog converters (DACs). The delays represent a phase shift in periodic signals that obstruct instantaneous feedback of the signals. Phase shift manipulation can be also utilized for time delay compensation as per the following explanation. Assuming that a 2nd order Butterworth low-pass filter with 200Hz cut-off frequency is used for noise filtering and anti-aliasing, the continuous Laplace domain transfer function that represents the filter plant, $TF_{LP}(s)$, is shown below

$$TF_{LP}(s) = \frac{0.0001551s^2 + 15.7s + 1.579e06}{s^2 + 1777s + 1.579e06} \tag{7}$$

The phase angle equation of a 2nd order Butterworth filter is determined using the following equation

$$\emptyset_{LP}(f_{exc}) = -\tan^{-1}\left[\frac{1}{\zeta}\left(2\frac{f_{exc}}{f_{LP}} + \sqrt{4-\zeta^2}\right)\right] - \tan^{-1}\left[\frac{1}{\zeta}\left(2\frac{f_{exc}}{f_{LP}} - \sqrt{4-\zeta^2}\right)\right] \tag{8}$$

where f_{LP}, f, and ζ are the cut-off frequency, the excitation frequency, and the damping ratio of the filter respectively. The frequency response of the above filter is plotted in **Figure** 7. The intention is to illustrate the induced phase shift between 45Hz and 55Hz.

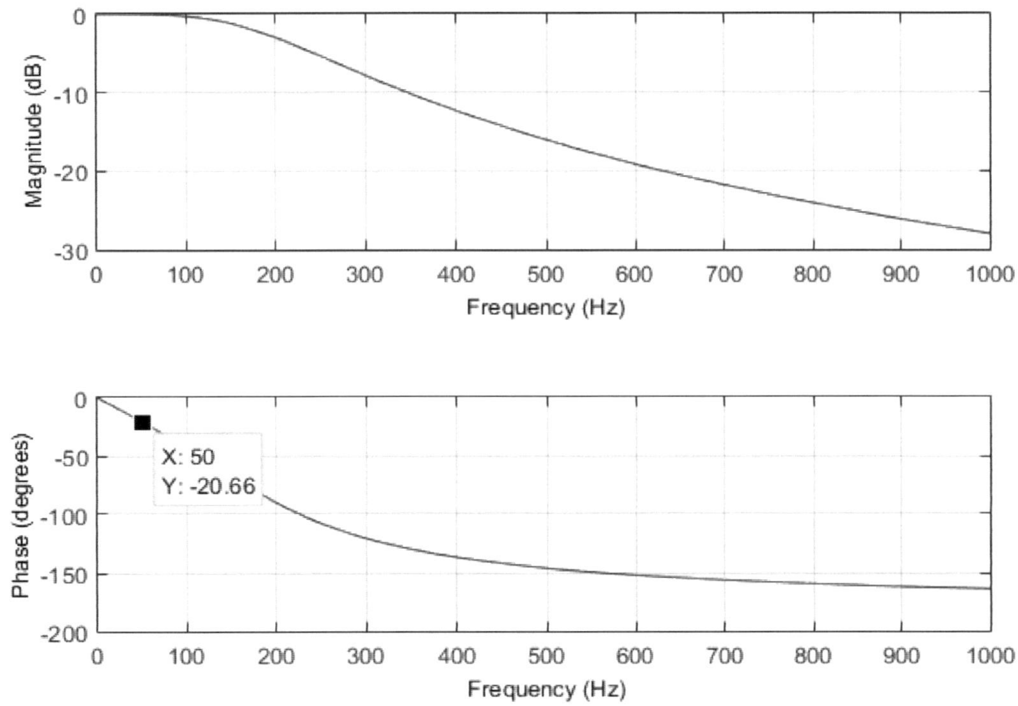

Figure 7. Low-pass Butterworth filter frequency response with cut-off frequency of 200Hz

Referring to the phase angle plot in the frequency range of 45-55Hz, the phase angles at 45Hz and 55Hz are -18.53^0 and -22.82^0 respectively. Those two phase shifts are associated with positive time delays of 1.143ms and 1.152ms at 45Hz and 55Hz respectively in accordance with the following relationship that associates time delay (τ) in second, the phase shift ($\emptyset$) in degrees, and the excitation frequency f_{exc}

$$\tau = \frac{\emptyset^0}{360^0 \times f_{exc}} \tag{9}$$

The phase shifts introduced by the low pass filter with 200Hz cut-off frequency have the effect of a time delay with a minimum value of 1.143ms at 45Hz and a maximum of 1.152ms at 55Hz.
Another kind of phase shift is now illustrated with the use of a 2nd order high-pass Butterworth filter with 1Hz cut-off frequency for the purpose of DC component elimination. The continuous Laplace domain transfer function that represents the filter plant, $TF_{Hp}(s)$, is shown below

$$TF_{HP}(s) = \frac{s^2 - 0.0003948s - 7.792e - 08}{s^2 + 8.886s + 39.48} \tag{10}$$

The phase angle equation of the 2nd order High-Pass Butterworth filter is determined using the following equation

$$\emptyset_{HP}(f_{exc}) = \pi - \tan^{-1}\left[\frac{1}{\zeta}\left(2\frac{f_{exc}}{f_{HP}} + \sqrt{4-\zeta^2}\right)\right] - \tan^{-1}\left[\frac{1}{\zeta}\left(2\frac{f_{exc}}{f_{HP}} - \sqrt{4-\zeta^2}\right)\right] \tag{11}$$

The phase equation for the high pass filter is similar to that of the low passes one advanced by a π radian. This means that while the Low-Pass filter corresponds to lagging the output signal, the High-Pass filter is associated with leading the output signal. The frequency response of the High-Pass filter used in the code is depicted in **Figure 8**.

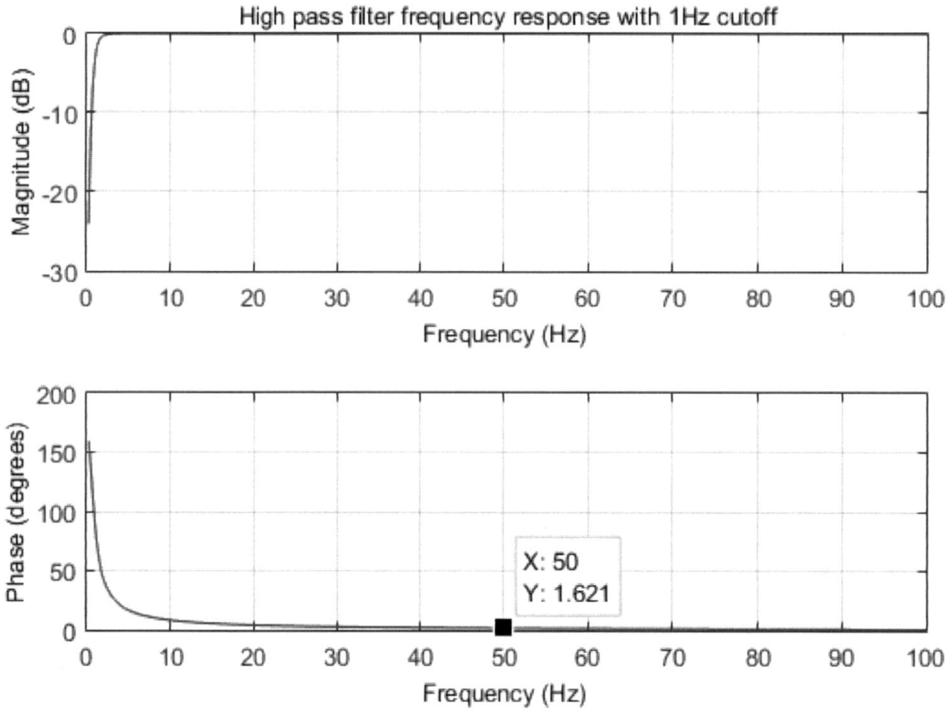

Figure 8. High-pass Butterworth filter frequency response with cut-off frequency of 1Hz

The phase angles at 45Hz and 55Hz are $1.801^0$and 1.473^0 respectively. Those phase shifts contribute to a max reverse time delayed output signal of -0.10006ms at 45Hz and min reverse time delay -0.07439ms at 55Hz. The group delay induced by the filtering stages is then estimated by adding the individual delays of each filter. Referring to the time delay margin identified in the investigation of the effect of time delays on control stability earlier, a careful choice of the cut-off frequencies of the low-pass and high-pass filters is capable of compensating the incurred group delays. The following discusses the derivation of the phase shift relationship between the High-Pass and Low-Pass filters cut-off frequencies. The aim is to obtain an analytic solution that relates the cut-off frequencies to the phase shift and then solve it to obtain a zero group phase shift. By adding the phase equations of the phase shift and equating to zero, the following equality is obtained

$$\frac{\zeta f_{exc}(f_{HP}+f_{LP})(-f_{exc}{}^2+f_{HP}f_{LP})}{\zeta^2 f_{exc}{}^2 f_{HP}f_{LP}-(f_{HP}f_{LP})^2+(f_{exc}f_{HP})^2+(f_{exc}f_{LP})^2-f_{exc}{}^2}=0 \qquad (12)$$

Assuming that cut-off frequency of one filter is known, the following shows the non-trivial solution that relates the excitation frequency to the cut-off frequencies

$$f_{exc}{}^2 = f_{LP}f_{HP} \qquad (13)$$

4. Implementation

This section is dedicated to the validation of the findings presented in the previous section. In the experiments, the shaker is powered by a 1.3 KVA power amplifier which is controlled by a small voltage signal between ±2 volts. The VCM (GVCM-051-051-01) is powered and controlled with a servo linear drive amplifier TA105 capable of delivering a max current of 2A peak to peak control by a fast proportional current controller. During operation, the active damping control method determines a reference current that is fed to the servo drive and then compared with the actual current in the VCM which is measured internally. An embedded CompactRio 9024 controller featuring an industrial 800

MHz real-time processor that runs LabVIEW for deterministic, reliable real-time applications is used for control, logging, and data analysis. The CRIO also features a user-defined FPGA circuitry in its chassis that controls each I/O module. Two LabVIEW codes were implemented simultaneously on the Compact Rio's real time engine and the FPGA respectively. The FPGA code is executed with a 20μs loop rate where the measurement and control takes place. The RT code runs with a 100ms loop rate and communicates to the FPGA code via direct memory access FIFOs. Two analogue input modules (NI 9234, NI 9205) and one analogue output module (NI 9263) were used to process the signals. Acceleration data of the shaker table and the TMD were obtained from two accelerometers with dynamic range of ±71g mounted in the middle of the shaker table and the helical spring respectively. The acceleration signals were sampled at a rate of 50KS/s using an AC coupled input stage of a 24-bit Analogue-to-Digital converter. A low-pass filter with a cut-off frequency of 200Hz is employed to remove any measurement noise from the accelerometers.

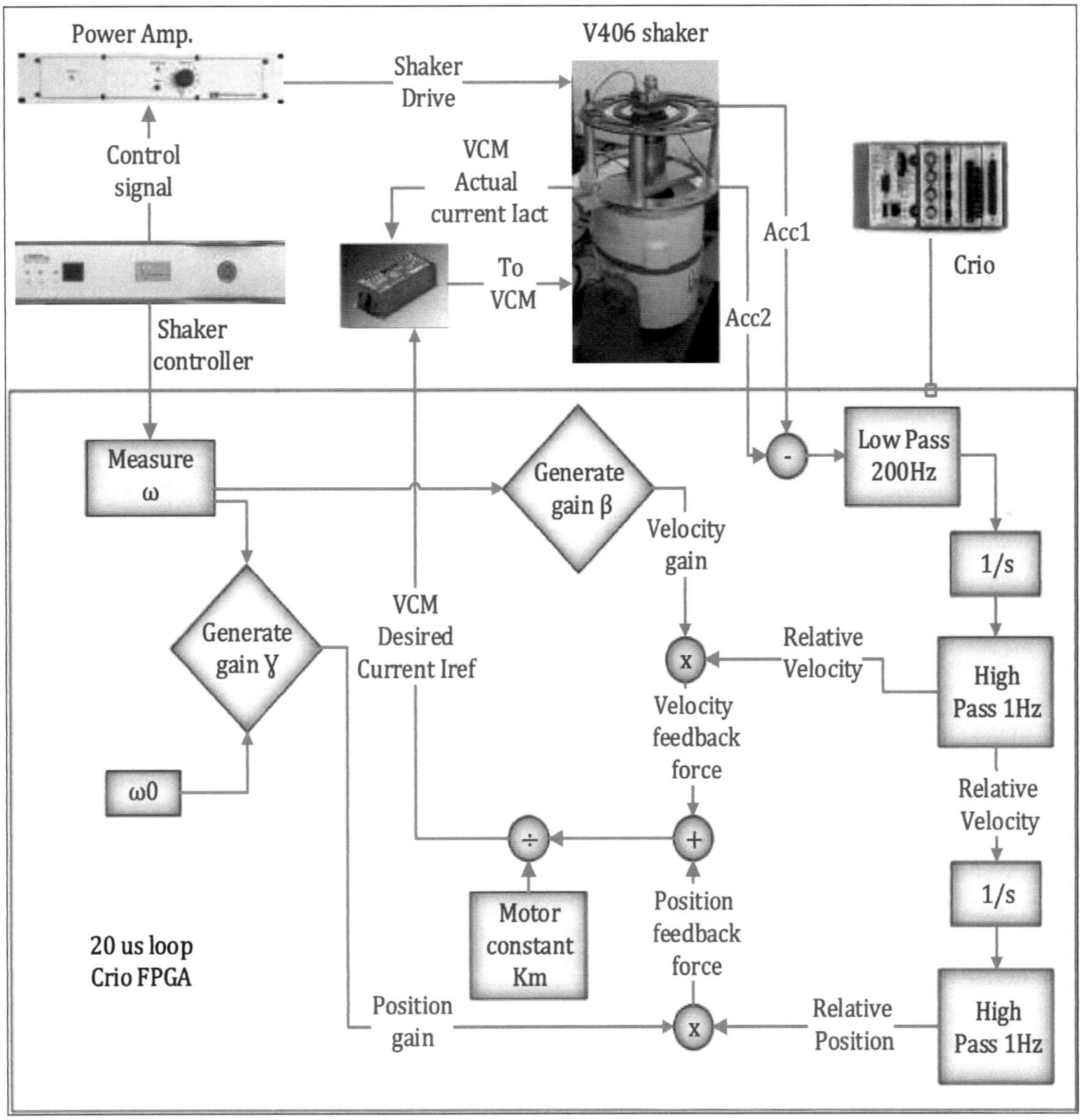

Figure 9. Schematic of the active damping control of Test 1 and Test 2

4.1. Test 1: Effect of induced time delays

In this test, the shaker excitation frequency is varied linearly from 40Hz to 60Hz in a time interval of 1 minute. The active vibration control is only operated in the frequency range between 45Hz and 55Hz. The ATMD is tuned at 50Hz and the active control aims to re-tune the ATMD frequency so it tracks that of the excitation and consequently cancel the vibration of the shaker table. Time delays of 1.143ms and 1.152ms are introduced between 45Hz and 55Hz by means of a second order Butterworth low-pass filter with 200Hz cut-off frequency that is originally used for noise and anti-aliasing. On the other hand, two High-Pass second orders Butterworth filters are implemented with a 1Hz cut-off frequency to remove any post-integration DC offset. The test parameters are summarized in **Table 2**. The value of the velocity feedback gain was chosen β = -14 Ns/m to remove the damping introduced from the VCM when the terminals are VCM are closed.

Table 2. Parameters for Test 1 and Test 2

Parameter	Symbol	Value	Unit
Shaker Excitation	V_s	$4\ \sin 2\pi f_{exc} t$	[V]
Excitation Frequency	f_{exc}	40 - 60	[Hz]
Test time	T	60	[s]
FPGA loop rate	Ts	20e-6	[s]
ATMD resonance	f_{ATMD}	50	[Hz]
Position feedback gain	γ	$4\pi^2 m_{\text{ATMD}}(f_{exc}{}^2 - f_{ATMD}{}^2)$	$[\text{Ns}^2/m]$
Velocity feedback gain	β	-14	$[Ns/m]$
VCM reference current	i_{ref}	$k_{vcm}{}^{-1}((\gamma(x_1 - x_2) + \beta(\dot{x}_1 - \dot{x}_2))$	[A]
"Control ON"	$f_{control}$	45 - 55	[Hz]
Low pass cut-off	f_{Lp}	200	
High pass cut-off	f_{Hp}	1	

The graphs of **Figure 10** present the experimental measurements of the shaker excitation and the shaker table acceleration for the test rig with passive TMD (black) and the actively controlled ATMD (red) for the purpose of comparison. The power consumption of the VCM is also plotted for the purpose of highlighting the efficiency of the control method.

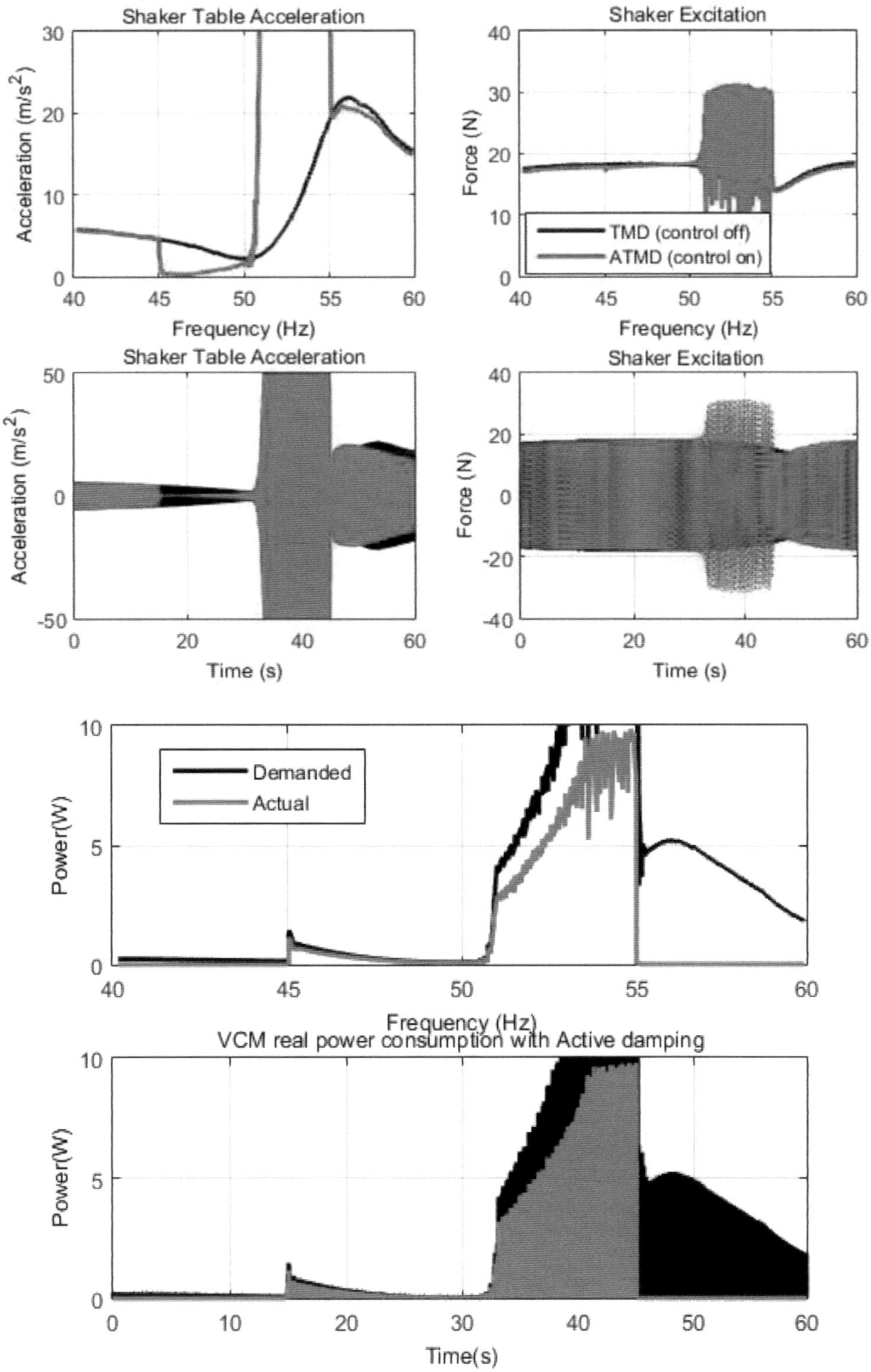

Figure 10. Experimental results of Test 1.

Referring to Figure 10, the top row of subplots on the left correspond to the magnitude of the shaker table acceleration and shaker force plotted against excitation frequency. The 2nd row of the subplots on the left shows the instantaneous acceleration and excitation force in against time. Note that the excitation frequency of the shaker is varied from 45Hz to 60Hz in 60 seconds. The figure on the right corresponds to the electrical power consumed by the VCM.

The shaker table acceleration with the passive case (black) is only attenuated at 50Hz which is the tuning frequency of the passive TMD. It is evident that the attenuation level doesn't cancel the vibration of the shaker table completely due to the damping in the VCM. The following discussion elaborates upon the active control (red) with ATMD whereby the response is analyzed in four different intervals.

1. [40-45]Hz: For the ATMD, the control is off in this interval and therefore the response of the shaker table acceleration is similar to that of the passive TMD response plotted in black. The current delivered to the VCM is zero in this interval due to the control being off. A small voltage appears across the terminals of the VCM in this interval due to the generated back EMF but this voltage is negligible.

2. [45-50]Hz: In the case of the passive TMD (black), the amplitude of the shaker table acceleration starts decreasing to reach a minimum at 50Hz due to resonance in the TMD. When control is on (red), the amplitude of the shaker table acceleration drops to minimum immediately at 45Hz due to the effect of the control action. As excitation frequency increases, it is revealed that the shaker table acceleration grows to larger amplitudes as soon as the excitation frequency approaches 50Hz. The maximum amplitudes power (1.8W) occurs at 45Hz. Theses amplitudes decrease gradually as the frequency reaches 50Hz.

3. [50-55]Hz: The control action is still on in this interval. The system becomes unstable after 50Hz and the amplitudes of the acceleration reach larger magnitudes as the frequency of excitation moves from 50Hz towards 55Hz. Similarly, the VCM voltage, current, and power reach larger values in this interval. The instability in the system caused higher demand in the VCM current due to the large acceleration amplitudes. This is evident in the graphs power where the actual power does not track the demanded ones. The source of instability in the control system is attributed to time delays resulting from filtering. The outcome of this test confirms the predictions obtained in the simulation earlier.

4. [50-55]Hz: This is the case when control is turned off in the system. No current is delivered to the VCM and hence the response of the shaker table acceleration is similar to that when the system is passive.

4.2. Test 2: Timed-delayed ATMD tuned at 55Hz

The theoretical findings of the effect of time delay on control stability were illustrated experimentally in implementation Test 1. Precisely, Test 1 showed that the control system becomes unstable when the excitation frequency is higher than the resonant frequency of the ATMD (50Hz) exactly similar to the observation in simulation results. It was also observed that the system only becomes unstable after the resonant frequency of the ATMD. Building upon this fact, one method of mitigating the effect of delays was proposed and simulated based on shifting the original resonance of the absorber up to 55Hz. The results of test 2 shown in **Figure 11** illustrate the success of the proposed method.
By examining the acceleration profile of the shaker table in **Figure 11** it is revealed that the active control succeeds in attenuating the vibration of the shaker table mass across the entire range of frequency range between 45Hz and 55Hz. The comparison between the power consumption of the VCM in this test with that of test 1 in **Error! Reference source not found.** shows that while this method is successful; more power is required than if the ATMD was originally tuned at 50Hz.

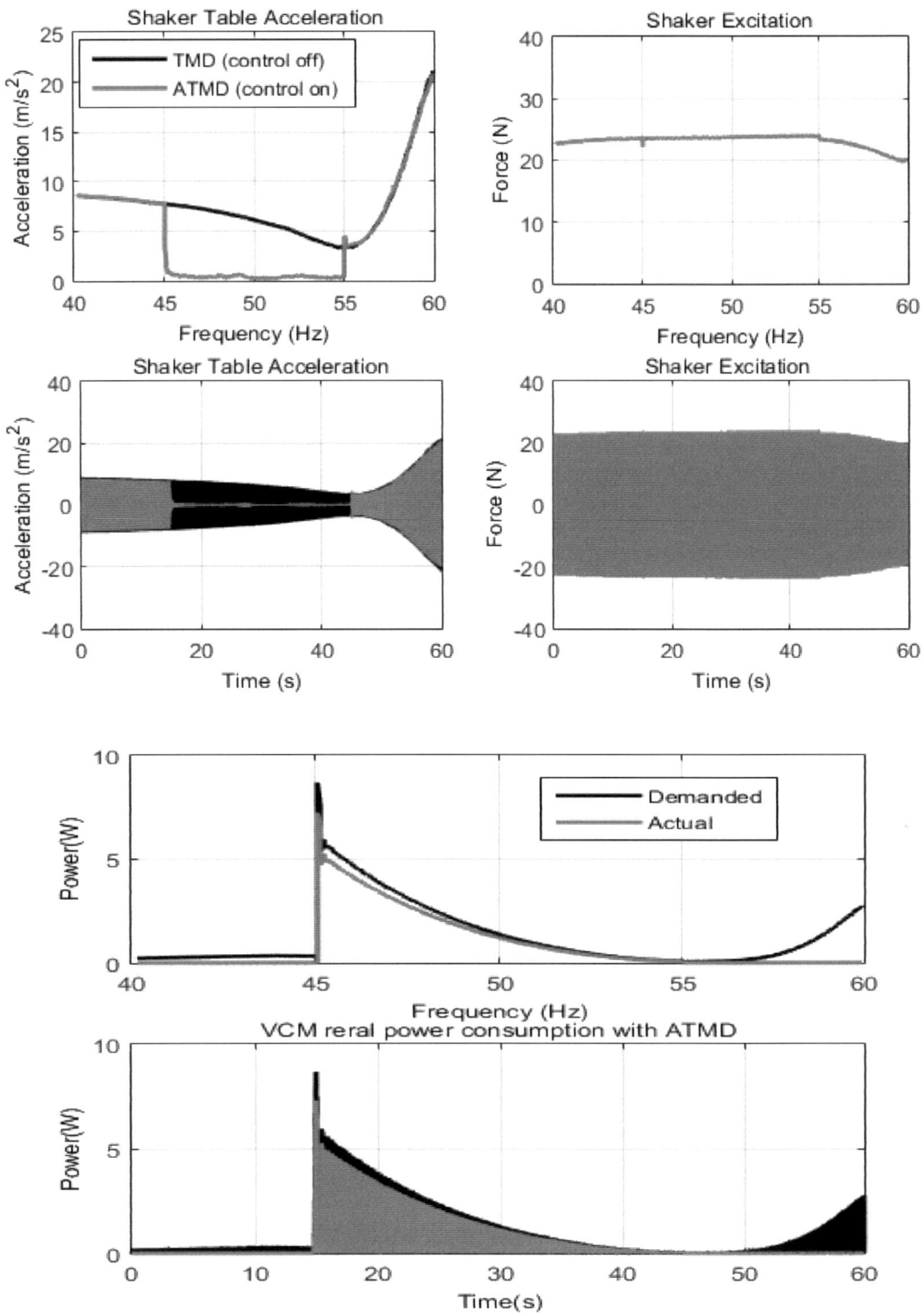

Figure 11. Experimental results of Test 2

4.3. Test 3: delay compensation

The compensation of time delays resulting from the filtering is achieved by adaptively altering the Low-Pass filter coefficients so that its cut-off frequency is chosen accordingly. The following schematic model represents that of Test 3.

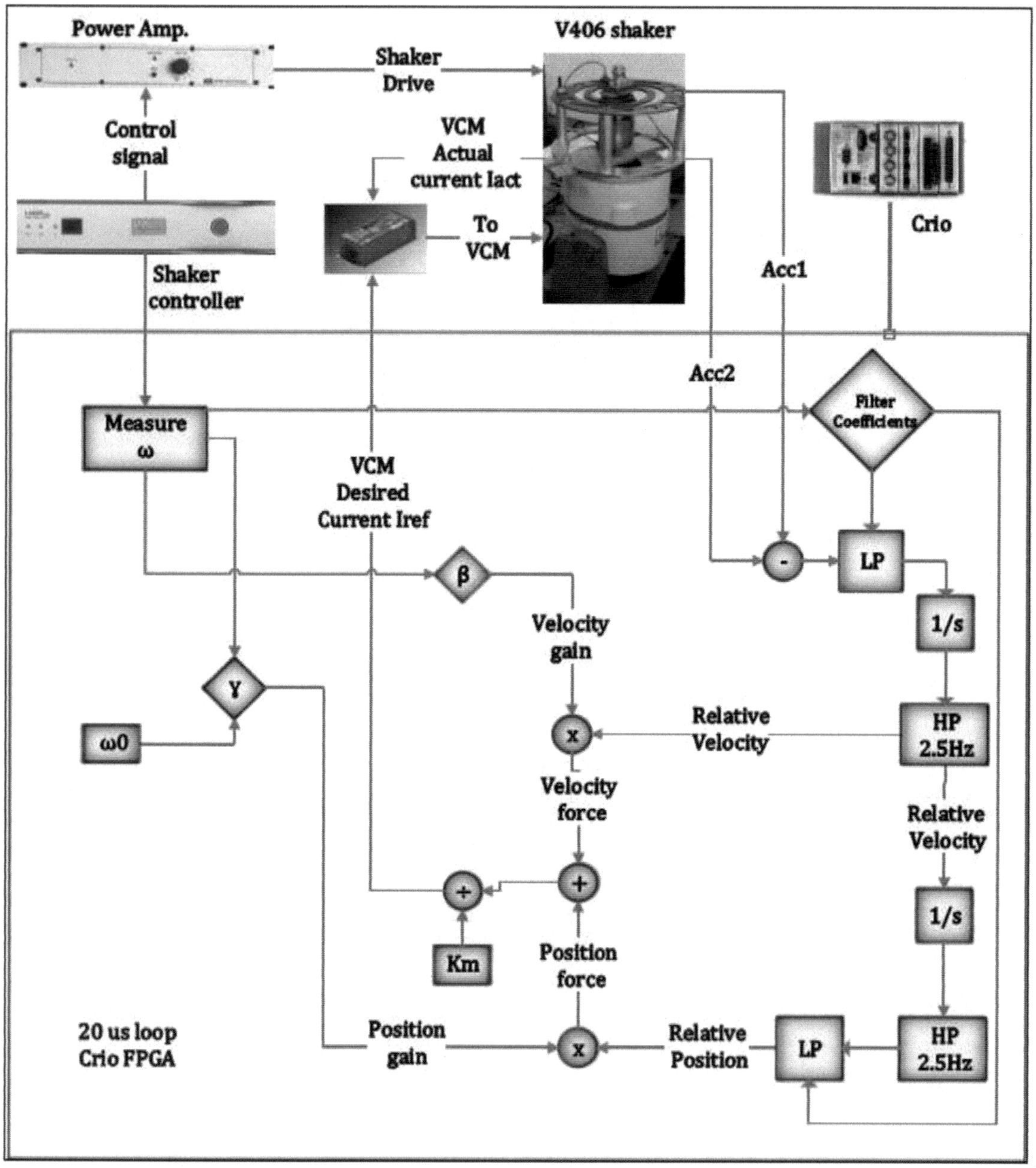

Figure 12. Schematic of the active damping control of Test 3

The values of **Table 3** are adopted in this experiment. The time and frequency domain results of this test are shown below.

Table 3. Parameters for Test 3

Parameter	Symbol	Value	Unit
Shaker Excitation	V_s	$4\ \sin 2\pi f_{exc}T$	[V]
Excitation Frequency	f_{exc}	40 - 60	[Hz]
Test time	T	60	[S]
FPGA loop rate	Ts	20e-6	[S]
ATMD resonance	f_{ATMD}	50	[Hz]
Position feedback gain	γ	$4\pi^2 m_{\mathrm{ATMD}}(f_{exc}{}^2 - f_{ATMD}{}^2)$	$\mathrm{N}s^2/m$
Velocity feedback gain	β	$-(0.103f_{exc}{}^2 - 9.58f_{exc} + 208)$	Ns/m
VCM reference current	i_{ref}	$k_{vcm}{}^{-1}((\gamma(x_1 - x_2) + \beta(\dot{x}_1 - \dot{x}_2))$	[A]
"Control ON"	$f_{control}$	45 - 55	[Hz]
Low pass cut-off	f_{Lp}	$f_{exc}{}^2/f_{HP}$	
High pass cut-off	f_{Hp}	2.5	

The test was performed while catering for the delays in the system as described above. For the shaker table acceleration, the active damping succeeded in attenuating its vibration between 45Hz and 55Hz with stable control action across the entire range.

By comparing the results with that of Test 1, the results of Test 3 are shown in **Figure 13**. According to the results, it is illustrated that delays were the source of instability encountered in Test 1. The compensation of time delays has successfully improved the system response by maintaining stability of the control after 50Hz. In Test 2, delays were mitigated by shifting the ATMD original resonant frequency to 55Hz so that it located outside the range of operation. The method was successful however it incurred extra power consumption as explained previously. Following this method, the electrical power consumption in the VCM is improved.

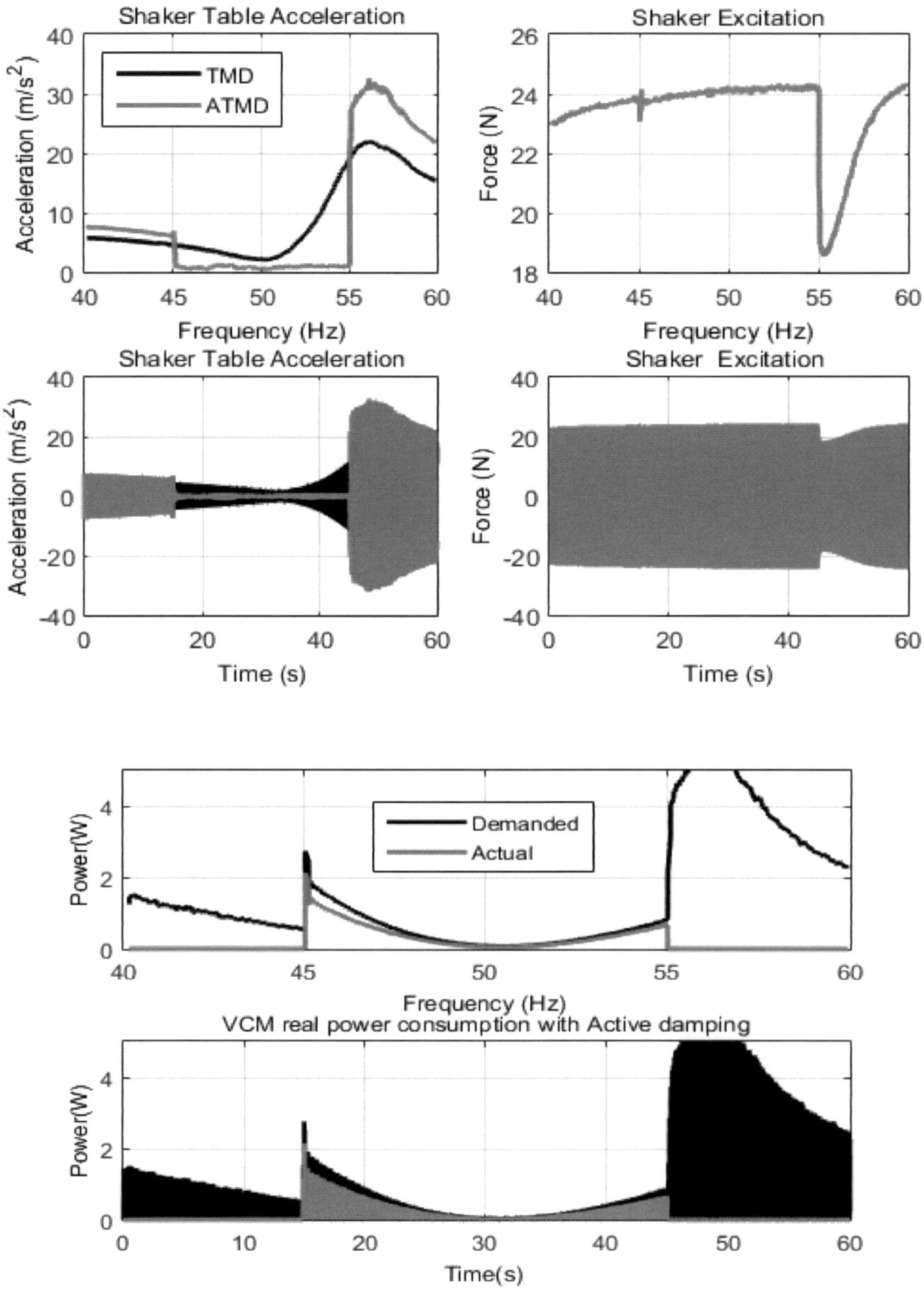

Figure 13. Experimental results of Test 3

5. Conclusion

In this work, an active vibration control method employing an electromagnetic ATMD with a voice coil actuator is employed for investigating a solution for the vibration problem in a Stirling machine with the aid of a scaled test rig. The experimental results with the test rig showed that the proposed active control law is capable of enhancing the frequency band of operation from 50±0.5Hz to between 45Hz-55Hz. A thorough theoretical and experimental investigation on the effect of time delays on control stability showed that after a critical time delay margin dependent of the excitation frequency the control system becomes unstable. Three methods for mitigating the effect of time delays on stability showed that it is possible to successfully mitigate the effect of time delays on the stability of the control system. The introduction of damping into that ATMD system improves the time delay margin allowing for more room of operation with a time-delayed ATMD. The second method incorporated re-tuning the original resonance of the ATMD which ensured stable control for excitation frequencies below the resonance of the ATMD. A third technique incorporating a dual-filtering technique of delay compensation provided a successful and less power-consuming solution compared to the other two. The proposed methods will not be only bound Stirling engine application specifically. In fact, it is useful for the compensation of the time delay effects in ATMD systems deployed in tall slender structures such as tall buildings, for effective mitigation of wind-induced vibrations.

References

[1] Walker G. 1980 *Stirling engines* Clarendon Press .

[2] Rao SS., 1990 *Mechanical vibrations* Addison-Wesley .

[3] Mead DJ. 1999 *Passive vibration control* Wiley .

[4] Ikeda Y. 2009 *Active and semi-active vibration control of buildings in Japan—Practical applications and verification* Structural Control and Health Monitoring 16 (7-8) 703-723 .

[5] De Roeck G, Degrande G, Lombaert G, Müller G. 2011 *A versatile active mass damper for structural vibration control* .

[6] Park K. Choi D. Ozer A. Kim S. Lee Y. Joo D. 2008 *A voice coil actuator driven active vibration isolation system with the consideration of flexible modes* Rev Sci Instrum 79 (6) 065106 .

[7] Liu Y. Wu W. 2013 *Active Vibration Isolation Using a Voice Coil Actuator with Absolute Velocity Feedback Control* International Journal of Automation and Smart Technology 3 (4) 221-226 .

[8] Qiu Z. Wu H. 2009 *Acceleration sensors based modal identification and active vibration control of flexible smart cantilever plate* Aerospace Science and Technology 13 (6) 277-290 .

[9] Thong YK. Woolfson MS. Crowe JA. Hayes-Gill BR. Jones DA. 2004 *Numerical double integration of acceleration measurements in noise* Measurement 36 (1) 73-92 .

[10] Liu B. Hu H. 2010 *Group delay induced instabilities and Hopf bifurcations, of a controlled double pendulum* Int J Non-Linear Mech 45 (4) 442-452 .

[11] Shao M. Chen W. 2012 *Active vibration control in a cantilever-like structure: a time delay compensation approach* J Vibrat Control 1077546312437802 .

[12] Jie Liu. Kefu Liu. and Hao Bai. 2005 *Application of a time-delayed control system in vibration suppression* Mechatronics and Automation, 2005 IEEE International Conference Anonymous .

[13] Cai G. Huang J. 2003 *Instantaneous optimal method for vibration control of linear sampled-data systems with time delay in control* J Sound Vibrat 262 (5) 1057-1071 .

[14] An F. Chen W. Shao M. 2014 *Dynamic behavior of time-delayed acceleration feedback controller for active vibration control of flexible structures* J Sound Vibrat 333 (20) 4789-4809 .

[15] Qiu Z. Han J. Zhang X. Wang Y. Wu Z. 2009 *Active vibration control of a flexible beam using a non-collocated acceleration sensor and piezoelectric patch actuator* J Sound Vibrat 326 (3–5) 438-455 .

[16] Hassan A. Torres-Perez A. Kaczmarczyk S. Picton P. 2015 *Active vibration control for a free piston Sirling engine generator using a voice coil actuator* mm (Modern Machinery) Science Journal (10.17973/MMSJ.2015_03_201503) 547 .

[17] Marques FA. Domingos A. Rade Sebastiao S. Cunha Jr. 2001 *Assessment of an active dynamic vibration absorber* Proceedings of IMAC-XIX: a conference on structural dynamics: February 5-8, 2001, Hyatt Orlando, Kissimmee, Florida Society for Experimental Mechanics Anonymous .

Appendix A

Coefficients of transfer function $W(s)$ shown in (6)

Numerator

$$b_5 = 0$$

$$b_4 = 0$$

$$b_3 = m_2 \, L_a$$

$$b_2 = -k_a^2 + c_2 \, k_p + c_2 \, R_a + k_2 \, L_a + \beta \, k_p - \gamma \, \tau \, k_p$$

$$b_1 = -k_a^2 + c_2 \, k_p + c_2 \, R_a + k_2 \, L_a + \beta \, k_p - \gamma \, \tau \, k_p$$

$$b_0 = k_2 \, k_p + k_2 \, R_a + \gamma \, k_p$$

Journal of Physics: Conference Series **721** (2016) 012007 doi:10.1088/1742-6596/721/1/012007

Denominator

$$a_7 = 0$$

$$a_6 = 0$$

$$a_5 = m_1 m_2 L_a$$

$$a_4 = k_p m_1 m_2 + c_1 m_2 L_a + c_2 m_1 L_a + c_2 m_2 L_a + m_1 m_2 R_a - \beta \tau k_p m_1 - \beta \tau k_p m_2$$

$$a_3 = \beta k_p m_1 - k_a^2 m_2 - k_a^2 m_1 + \beta k_p m_2 + c_1 k_p m_2 + c_2 k_p m_1 + c_2 k_p m_2 + c_1 c_2 L_a + c_1 m_2 R_a + c_2 m_1 R_a + c_2 m_2 R_a + k_1 m_2 L_a + k_2 m_1 L_a + k_2 m_2 L_a$$
$$- \beta \tau c_1 k_p - \gamma \tau k_p m_1 - \gamma \tau k_p m_2$$

$$a_2 = -c_1 k_a^2 + \beta c_1 k_p + \gamma k_p m_1 + \gamma k_p m_2 + c_1 c_2 k_p + c_1 c_2 R_a + k_1 k_p m_2 + k_2 k_p m_1 + k_2 k_p m_2 + c_1 k_2 L_a + c_2 k_1 L_a + k_1 m_2 R_a + k_2 m_1 R_a + k_2 m_2 R_a$$
$$- \gamma \tau c_1 k_p - \beta \tau k_1 k_p$$

$$a_1 = -k_1 k_a^2 + \gamma c_1 k_p + \beta k_1 k_p + c_1 k_2 k_p + c_2 k_1 k_p + c_1 k_2 R_a + c_2 k_1 R_a + k_1 k_2 L_a - \gamma \tau k_1 k_p$$

$$a_0 = \gamma k_1 k_p + k_1 k_2 k_p + k_1 k_2 R_a$$

Simple models for rope substructure mechanics: Application to electro-mechanical lifts

I Herrera[1a], **S Kaczmarczyk**[2b]

[a]Escuela de Ingenierías Industriales, Universidad de Extremadura, Avda. de Elvas s/n, 06006 Badajoz, Spain

[b]School of Science and Technology, The University of Northampton, St. George's Avenue, Northampton, NN2 6JD, UK

[1]iherrera@unex.es, [2]stefan.kaczmarczyk@northampton.ac.uk

Abstract. Mechanical systems modelled as rigid mass elements connected by tensioned slender structural members such as ropes and cables represent quite common substructures used in lift engineering and hoisting applications. Special interest is devoted by engineers and researchers to the vibratory response of such systems for optimum performance and durability. This paper presents simplified models that can be employed to determine the natural frequencies of systems having substructures of two rigid masses constrained by tensioned rope/cable elements. The exact solution for free un-damped longitudinal displacement response is discussed in the context of simple two-degree-of-freedom models. The results are compared and the influence of characteristics parameters such as the ratio of the average mass of the two rigid masses with respect to the rope mass and the deviation ratio of the two rigid masses with respect to the average mass is analyzed. This analysis gives criteria for the application of such simplified models in complex elevator and hoisting system configurations.

1. Introduction

We focuses our attention on the mechanical substructure shown in figure 1 in which two lumped mass components M_1 and M_2 are connected by an elastic tensioned member/rope-piece of length l which can be the equivalent mechanical model of a number of common substructures in lift engineering and hoisting equipment. We are particularly interested in traction lift suspension ropes substructures. It should be mention that the system in figure 1 is a semi-definite system. Thus, there is a rigid body mode with zero frequency.

The prediction of dynamic response of the suspension ropes is essential to understand and model the elevator system. In a typical lift installation the car/counterweight suspension system is formed by a number (n_R) of equally tensioned steel wire ropes acting in parallel, attached to the car frame at one end, passing over a traction sheave and diverter pulleys to the counterweight at the other end. The contact between the rope and traction sheave and diverter pulleys is specially critical, not only for the neccesity to produce the desired motion of the car but for its control, safety, and maintenance of relevant traction condition according to EN-81-20. If we assume that no slipping exists between the ropes and the traction sheave both sides of the rope can be represented by the simplified substructure shown in figure 1. Thus: M_1 would correspond to the traction sheave equivalent mass (rotary or linear) for both sides and M_2 would correspond to the car mass for the frame side substructure or the counterweight mass for the counterweight side substructure. Another model from the opposite side to that free (M_1)-free (M_2) model shown in figure 1 is the fixed (M_1)-free (M_2) model which was extensively studied by Leung [2]. This model could be the best aproximation for static stopped elevator slings(brakes pressed) but its response wouldn't be affected at all by the lumped mass (M_1).

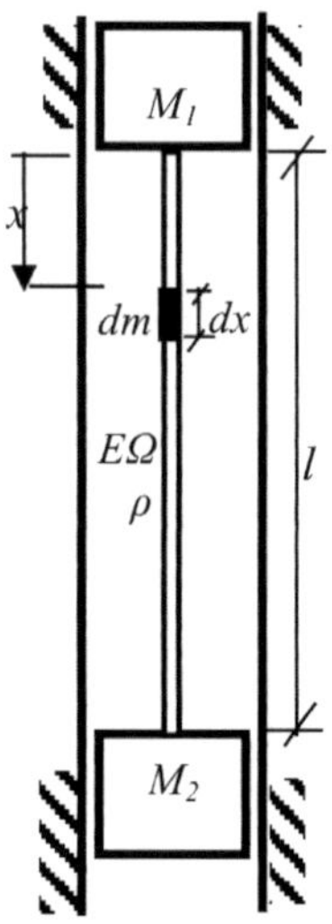

Figure 1. Substructure of two rigid masses tensioned by a rope.

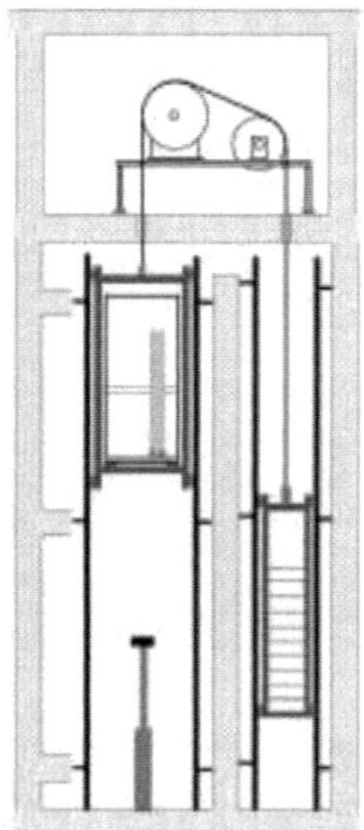

Figure 2. Elevator system schem.

The dynamic behavior of the mass suspension system can be analyzed by treating the ropes as linear distributed-parameter elastic members. To give criteria about the relative influence of the mechanical characteristics of the rope-pieces with respect to the joined up rigid masses properties and the method of analysis, several substructuring techniques were studied [2-3] and the system shown in figure 1 will be analyzed.

The dynamics of suspended ropes has been intensively studied. Several contributions from Zhu et al. [4-5] and Kaczmarczyk [6-7] give a wide and complete solution to the rope modelled as linear elastic continua. The assumptions for the car and counterweight as assemblies to be treated as rigid bodies are based on the low longitudinal elastic modulus of the rope whilst perfect flexural flexibility and high elastic modulus of the metallic components.

Typical analysis capable of generalization and modularity of complex systems is the application of the finite element method [8-9]. The continuous description of the rope is substituted by a finite number of sections/ elements, with their deformations described by equations in terms of unknowns displacements.

In the simplified models it is proposed to discretize the suspended rope- substructure using only one element. This has the advantage of maintaining small number of unknowns and equations and facilitates any further synthesizing strategy. In such an approach these models describe the elevator response well without significant loss of accuracy. Particularly, special interest is devoted to justify the application to suspended rope-substructures which were discretized to adopt a 6 degrees of freedom discrete model of the elevator system [1].

In a typical lift or elevator system the car and the counterweight are mounted in a frame/sling (see figure 2). Both are generally made of steel sections and fixed to each counter-part by bolts and nuts or by welding techniques [1]. Special attention is devoted to tighten the pieces to avoid noise during travel. Each of these components can be treated as an ideal rigid body constrained by the lift guiding system. In this arrangement the six degrees of freedom of a rigid body moving in space are reduced to only one (vertical translation) assuming ideal sliding joints between car-frame or counterweight and the guide rails.

2. Boundary-value problem solutions

Two rigid components of masses M_l and M_2 are connected by an elastic tensioned rope-piece of length l (figure 1). The rope-piece is characterized by its density ρ, net cross-section area Ω and apparent elastic modulus E and then its total mass m ($m=\rho\Omega l$). Let us denote by M the average mass of M_l and M_2. Then, two characteristics ratios of the system are intended to be analyzed: the ratio of the average rigid masses M with respect to the rope mass m, M/m and the ratio of the semi-difference mass $(M_l-M_2)/2$ with respect to the average mass, d:

$$d=\frac{\frac{M_l-M_2}{2}}{M}\Rightarrow M_l=M(1+d);M_2=M(1-d) \tag{1}$$

The longitudinal vibrations (u) of the rope-piece are characterized by the differential equation [10]:

$$\ddot{u}-b^2u''=0 \tag{2}$$

where the dot denotes the differentiation with respect to the time t and the prime denotes derivative with respect to x, and $b=\sqrt{E/\rho}$. The solution of (1) for stationary vibrations can be determined if u has the form $u=X(x)\cdot T(t)$ where X is a function of x only and T is a function of t only. Substituting this in (2) gives

$$b^2\cdot X''\cdot T=X\cdot\ddot{T} \tag{3}$$

By separating the variables,

$$T=A\cos(\omega t+\theta);X=C\sin\frac{\omega}{b}x+D\cos\frac{\omega}{b}x \tag{4}$$

where A, B and C are integration constants. Combining the constants yields:

$$u=\left(\bar{C}\sin\frac{\omega}{b}x+\bar{D}\cos\frac{\omega}{b}x\right)\cos(\omega t+\theta) \tag{5}$$

where $\bar{C}=AD$ and $\bar{D}=AD$.

The constants $\bar{C}$ and $\bar{D}$ are determined by the boundary conditions of the rope-piece and by the initial conditions of the vibration. The boundary conditions are:

$$\begin{aligned}&at\ x=0:\ E\Omega\cdot u'(0,t)=M_1\cdot\ddot{u}(0,t)\\&at\ x=l:\ E\Omega\cdot u'(l,t)=-M_2\cdot\ddot{u}(l,t)\end{aligned} \tag{6}$$

which leads to:

$$\begin{aligned}&E\Omega\frac{\omega}{b}\bar{C}\cdot\cos(\omega t+\theta)=M_1\omega^2\bar{D}\left[-\cos(\omega t+\theta)\right]\\&E\Omega\frac{\omega}{b}\left(\bar{C}\cos\frac{\omega l}{b}-\bar{D}\sin\frac{\omega l}{b}\right)\cdot\cos(\omega t+\theta)=-M_2\omega^2\left(\bar{C}\sin\frac{\omega l}{b}+\bar{D}\cos\frac{\omega l}{b}\right)\left[-\cos(\omega t+\theta)\right]\end{aligned} \tag{7}$$

This set of equations is compatible and undetermined if:

$$tg\alpha = \frac{\alpha\left(\frac{M_1}{m}+\frac{M_2}{m}\right)}{\alpha^2\frac{M_1}{m}\frac{M_2}{m}-1} \tag{8}$$

where $\alpha=\omega l/b$. The solution of equation (8) gives n eigenvalues α_i $(i=1,\dots,n)$. Every α_i defines the frequency of the system ω_i in terms of the parameters M_1/m, M_2/m, l and b, so that the complete solution of (2) is of the form:

$$u=\sum_{i=1}^{i=\infty}\left(\overline{C}_i \sin\frac{\omega_i}{b}x+\overline{D}_i \cos\frac{\omega_i}{b}x\right)\cos(\omega_i t+\theta_i) \tag{9}$$

Equation (8), which can be re-written in terms of parameters M and d as Equation (10), gives the eigenvale α_1 corresponding to the fundamental frequency ω_1 ($\omega_1=\alpha_1 b/l$).

$$tg\alpha = \frac{\alpha\left(\frac{2M}{m}\right)}{\alpha^2\frac{(M-dM)}{m}\frac{(M+dM)}{m}-1} = \frac{2\alpha\frac{M}{m}}{\alpha^2\left(\frac{M}{m}\right)^2\left(1-d^2\right)-1} \tag{10}$$

The eigenvalue α_1 which corresponds to the lowest root of (10) is listed for a number of mass M/m and d ratios in the table 1. It is evident that the first frequency of the substructure will be influenced by both the mass ratio and the deviation ratio. The limitation of this exact solution is the difficulty to be extended to damped and non steady state solutions. An alternative to the exact solution based on the frequency equation (10) capable of increasing complexity and modularity is that based on the finite element method (FEM).

Table 1. First frequency factor α_1 for a number of M/m and d ratios.

M/m	$d=0$	$d=0.3$	$d=0.5$
0.0100	3.0800	3.0800	3.0800
0.0200	3.0209	3.0209	3.0210
0.0300	2.9642	2.9643	2.9645
0.0400	2.9098	2.9101	2.9106
0.0500	2.8577	2.8582	2.8590
0.0600	2.8078	2.8085	2.8098
0.0700	2.7599	2.7610	2.7628
0.0800	2.7140	2.7154	2.7180
0.0900	2.6699	2.6718	2.6752
0.1000	2.6277	2.6301	2.6343
0.2000	2.2845	2.2933	2.3089
0.3000	2.0422	2.0575	2.0849
0.4000	1.8615	1.8820	1.9193
0.5000	1.7207	1.7451	1.7902
1.0000	1.3065	1.3391	1.4028
2.0000	0.9602	0.9931	1.0605
5.0000	0.6221	0.6482	0.7039
10.0000	0.4435	0.4635	0.5067

In the FEM approach the rope-piece is subdivided into a number of finite axially loaded bar elements of length l_e and linear interpolation for the spatial part $X(x)$ of the unknown displacement u is proposed which gives the oportunity to include the influence of damping, essential for modelling real elevator systems. This drives to a finite consistent element with two degrees of freedom characterized dynamically [12] by the matrices *[m_e], [k_e]* and *[c_e]* as follows:

$$[m]^e = m\begin{bmatrix} \frac{1}{3} & \frac{1}{6} \\ \frac{1}{6} & \frac{1}{3} \end{bmatrix};[k]^e = \frac{E\Omega}{l^e}\begin{bmatrix} 1 & -1 \\ -1 & 1 \end{bmatrix};[c]^e = \frac{c}{l^e}\begin{bmatrix} 1 & -1 \\ -1 & 1 \end{bmatrix} \tag{11}$$

where c is the viscous damping coefficient in the rope that is assigned depending of the suspended overall mass and will be defined further on. If we discretize the rope-piece shown in Figure 1 by only one element following the Finite Element method, assemble together to the concentrated masses M_1 and M_2 and express the governing equations using the matrices of the system *[M], [K]* and *[C]*, force vector *[F]* and unknown displacements *[u]* we get:

$$[M][\ddot{u}]+[C][\dot{u}]+[K][u]=[F] \tag{12}$$

where:

$$[M] = \begin{bmatrix} M_1+\frac{m}{3} & \frac{m}{6} \\ \frac{m}{6} & M_2+\frac{m}{3} \end{bmatrix};[K] = \frac{E\Omega}{l}\begin{bmatrix} 1 & -1 \\ -1 & 1 \end{bmatrix};[C] = \frac{c}{l}\begin{bmatrix} 1 & -1 \\ -1 & 1 \end{bmatrix} \tag{13}$$

Another coarse simplified model for the dynamics shown in Figure 1 is based on the discrete approach in which the total mass of the rope is redistributed into two concentrated masses on its ends and these two masses are connected with and ideal equivalent spring/damper. Then we get the matrices as follows:

$$[M] = \begin{bmatrix} M_1+\frac{m}{2} & 0 \\ 0 & M_2+\frac{m}{2} \end{bmatrix};[K] = \frac{E\Omega}{l}\begin{bmatrix} 1 & -1 \\ -1 & 1 \end{bmatrix};[C] = \frac{c}{l}\begin{bmatrix} 1 & -1 \\ -1 & 1 \end{bmatrix} \tag{14}$$

Finally, if we consider the rope massless, the matrices of the governing equations are:

$$[M] = \begin{bmatrix} M_1 & 0 \\ 0 & M_2 \end{bmatrix};[K] = \frac{E\Omega}{l}\begin{bmatrix} 1 & -1 \\ -1 & 1 \end{bmatrix};[C] = \frac{c}{l}\begin{bmatrix} 1 & -1 \\ -1 & 1 \end{bmatrix} \tag{15}$$

We can compare these simplified models of the substructure shown in Figure 1 with the exact solution solving:

$$\left| -\omega^2[M]+[K] \right| = 0 \tag{16}$$

which is the condition for getting the natural frequencies of the undamped system. The eigenvalue problem represented by Equation (16) can be easily solved using established numerical techniques (implemented in software tools such as the '*eig*' MATLAB function). The analytical solution for the massless rope model gives the following fundamental frequency:

$$\omega_1 = \sqrt{\frac{2}{1-d^2}} \cdot \sqrt{\frac{E\Omega/l}{M}} \tag{17}$$

which is independent of the mass ratio *M/m*.

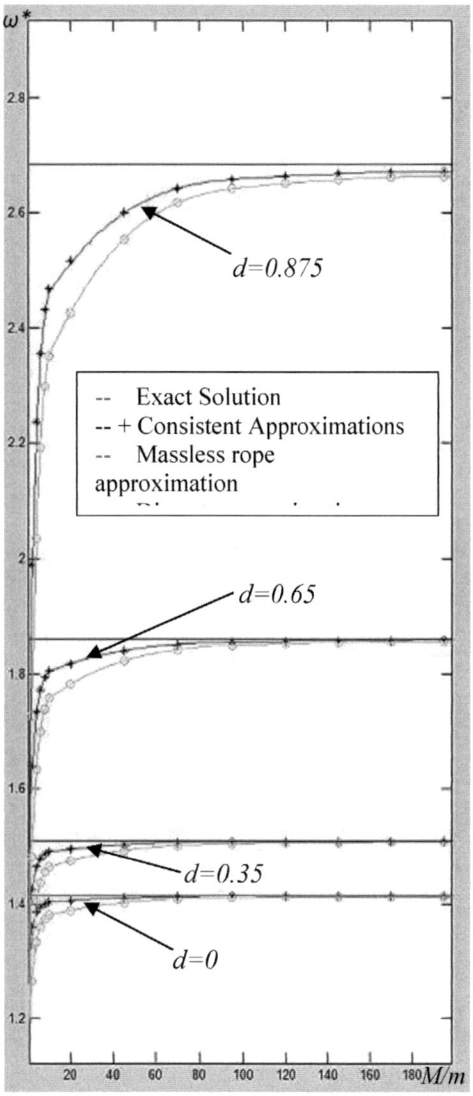

Figure 3. The first frequency dimensionless factor ω^* of the substructure shown in figure 1 against the mass ratio *M/m* for a number of deviation ratios *d* and models.

It can be shown that there is a simple closed-form formula for the models (14) and (15)

$$\omega_1 = \sqrt{\frac{k}{m_{eff}}} \tag{18}$$

where $k = E\Omega/l$ and $m_{eff} = \dfrac{m_1 m_2}{m_1 + m_2}$ where m_i=M_i, i=1, 2 for model (15) and m_i=M_i+$m/2$, i=1, 2 for model (14).

3. Results and Conclusions

Figure 3 shows the dimensionless parameter ω^* defined which is related to the first frequency of the substructure ω_1 (ω_1=$\alpha_1 b/l$), as a function of the mass ratio M/m for a number of deviation ratios.

$$\omega^* = \frac{\omega_1}{\sqrt{\dfrac{E\Omega/l}{M}}} = \frac{\omega_1}{\sqrt{\dfrac{E\Omega/l}{m}}}\sqrt{\frac{M}{m}} = \alpha_1\left(\frac{M}{m}, d\right)\sqrt{\frac{M}{m}} \tag{19}$$

It is observed how the consistent simplification gives the best approximation in practically every condition of mass and deviation ratios with respect to the other simplification models. Considering the rope massless -equations (15)- or using the discrete aproximation –equations (14)- gives a considerable error in the range of mass ratio which applies to elevator applications (M/m=2 and d=0: α_1=0.9602 and ω^*=0.9602*(2)^0.5=1.3579 and by (19) ω^*_{aprox}= (2)^0.5, that is, an *3.98%* error), though it can be acceptable for coarse approximations.

We have studied the consistent approximation for a number of typical configurations of residential elevators (number of passengers n_P ranging from 4 to 16) in the car-frame sling which is worse case than that of the counterweight sling. We have considered both the rotary and the translational lift drive-car frame substructures (See table 2). The mass of the sling m' has been computed according to the number of ropes n_R and roping system (*n:1*). It is observed that the mass deviation d is always below 75.1% and the maximum mass ratio is above 1.46 (M/m>1.46) giving an error in the fundamental frequency in the range *0.3%-1.4%* after computing equation (10) and solving equation (16). The question that must be discussed now is whether the same results would arise by analyzing higher natural frequencies of the substructure.

Table 2. Application to car-frame sling of residential elevators

n_P	M_2 Car-frame (empty car) [kg]	M_1 Lift Drive +bedplate [kg]	M_1=0.75·I/R^2 Rotary lift drive [kg]	n_R(n:1)	ρ [kg/m]	m'=ρhnn$_R$ (h=30m) [kg]	M [kg]	d	M/m'
4	167	245	400x72 : 53.97	3(1:1)	0.423	38.07	110	0.511	2.90
		203	240x110: 29.69	4(2:1)	0.179	42.96	98	0.698	2.29
6	167	310	400x87: 65.22	4(1:1)	0.423	50.76	116	0.438	2.29
		220	240x110 29.69	5(2.1)	0.179	53.7	98	0.698	1.83
8	167	300+50	400x102: 76.46	6(1:1)	0.423	76.14	122	0.372	1.60
		182+55	240x110 29.69	6(2:1)	0.179	64.4	98	0.698	1.53
13	241	350+75	400x72: 53.97	6(1:1)	0.423	76.14	148	0.634	1.94
		214+70	240x130 35.08	8(2:1)	0.179	85.92	138	0.746	1.61
16	247	450+100	400x102: 76.46	6(1:1)	0.610	109.8	161	0.527	1.47
		232+80	240x130: 35.08	9(2:1)	0.179	96.66	141	0.751	1.46

This was studied by Leung in the case of longitudinal vibrations in a fixed-free bar [2]. It is obvious that p natural frequencies can only be predicted by a p degree-of-freedom mathematical model. It is evident that the accuracy of the predicted frequencies decreases with the increase of the number of degrees of freedom.

The consistent simplification is the best one element-substructure. The application of simple models to suspended rope-substructures of only 6 degrees of freedom discrete model of an elevator system seems to be adequate and coherent. In future work, the effect of increasing the number of elements of the rope-piece using the FEM method for increasing knowledge in the dynamic response of the elevator system, specially to higher frequencies and retardation estimations, must be analyzed.

Acknowledgements
Authors would like to thank Mr. José Luis Yus from MP Ascensores for his help with the compilation of weight distribution data of commercial elevators.

References

[1] Sicilia J R, Alonso F J and Herrera I 2015 Mechanical Discrete Simulator of the Electromechanical Lift with n:1 roping *The 5th Symposium on Mechanics of Slender Structures (MoSS2015)* 21-22 September 2015 Northampton, UK.
[2] Leung A Y T 1993 *Dynamic Stiffness and Substructures* Springer-Verlag London
[3] Craig RR 1995 Substructure Methods in Vibration *Transactions of the ASME* **117**
[4] WD Zhu, H Ren An Accurate Spatial Discretization and Substructure Method With Application to Moving Elevator Cable-Car Systems—Part I: Methodology *Journal of Vibration and Acoustics* october 2013 **135**
[5] Zhu WD and Ren H 2013 An Accurate Spatial Discretization and Substructure Method With Application to Moving Elevator Cable-Car Systems—Part II: Application. *Journal of Vibration and Acoustics* october 2013 **135**
[6] Kaczmarczyk S and Mirhadizadeh S 2014 Modelling, Simulation and Experimental Validation of Nonlinear Dynamic Interactions in an Aramid Rope System *Applied Mechanics and Materials Special Issue: Dynamics and Control of Technical Systems* **706** 117-127 1662-7482
[7] Kaczmarczyk S and Picton P D 2013 The Prediction of Nonlinear Responses and Active Stiffness Control of Moving Slender Continua Subjected to Dynamic Loadings in Vertical Host Structures *The International Journal of Acoustics and Vibration* **18**(1) 39–44
[8] Stokey W F 2002 Vibration of systems having distributed mass and elasticity *Shock and Vibration Handbook* Chapter 7. Fifth Edition. Cyril M. Harris. Mc Graw Hill.
[9] Alarcón E 1990 Cálculo Dinámico Chapter 4 *Cálculo Matricial de Estructuras* Editorial Reverté.
[10] Belluzi O 1971 *Ciencia de la Construcción IV*. 443-445 Editorial Aguilar

Software design to calculate and simulate the mechanical response of electromechanical lifts

I Herrera[1a], E Romero[2a]

[a]Escuela de Ingenierías Industriales, Universidad de Extremadura, Avda. de Elvas s/n, 06006 Badajoz, Spain

[1]iherrera@unex.es, [2]eromero@unex.es

Abstract. Lift engineers and lift companies which are involved in the design process of new products or in the research and development of improved components demand a predictive tool of the lift slender system response before testing expensive prototypes. A method for solving the movement of any specified lift system by means of a computer program is presented. The mechanical response of the lift operating in a user defined installation and configuration, for a given excitation and other configuration parameters of real electric motors and its control system, is derived. A mechanical model with 6 degrees of freedom is used. The governing equations are integrated step by step through the Meden-Kutta algorithm in the MATLAB platform. Input data consists on the set point speed for a standard trip and the control parameters of a number of controllers and lift drive machines. The computer program computes and plots very accurately the vertical displacement, velocity, instantaneous acceleration and jerk time histories of the car, counterweight, frame, passengers/loads and lift drive in a standard trip between any two floors of the desired installation. The resulting torque, rope tension and deviation of the velocity plot with respect to the setpoint speed are shown. The software design is implemented in a demo release of the computer program called ElevaCAD. Further on, the program offers the possibility to select the configuration of the lift system and the performance parameters of each component. In addition to the overall system response, detailed information of transients, vibrations of the lift components, ride quality levels, modal analysis and frequency spectrum (FFT) are plotted.

1. Introduction

Design engineers are interested in the response of the elevator system in a number of circumstances. The most typical for that is the normal functioning in a standard trip like quantifying car position, acceleration, speed or jerk, assess ride quality, roller guides efficiency or levelling problems though other exceptional circumstances like impacts provided by guide misalignment, buffer actuation or safety gear actuation by means of uncontrolled car movements are of interest too.

A mechanical model with 6 degrees of freedom of the elevator system connected by generalized joints was developed and the governing equations for residential or freight lifts with any roping arrangement were derived [1]. This paper presents the analysis, design and programming of software that enables the calculation of the mechanical response of a real electromechanical lift installation.

Several solutions strategies of the governing equations are analyzed in order to get the mechanical response of the system in every scenario. The main, more versatile, strategy consists on a step by step integration using the Kutta-Meden algorithm which can be implemented in the MATLAB platform.

The expected response consists of various variables. Firstly, the response includes the vertical displacement, velocity and acceleration of the car, counterweight, frame, passengers/loads and lift drive as a function of time. Secondly, the variables related to the electric and electronic drives, mainly the deviation of the lift drive speed from the setpoint speed which is used for the electronic controller to determine the machine supply frequency according to a predefined control law, and the torque applied to the machine.

Vertical transport control involves the conjunction of a number of components of different nature. Each of these components has its own dynamic behaviour that could be modelled by differential equations. In this paper, it is analysed the mechanical dynamics considering that both, electronic and electric components, have an immediate response in comparison with mechanical component dynamics, and are modelled by static equilibrium equations instead of using the corresponding differential equations. The main objective of including these static models within the general mechanical differential model is to highlight the differences when using different electronic controllers or electrical machines.

2. Input parameters and initial conditions

The characteristic of a certain installation are specified by means of the parameters shown in the right of figure 1 where h^i represents the distance from the *0*-floor landing to the *i-th* floor landing and the colored circles represents the local origin of the displacements masses: red for the counterweight (y_{CW}), black for the car frame (y_{FR}), yellow for the car (y_{CA}), green for the passengers/loads (y_{PA}), blue for the machine drive (y_{MA}) and cyan for the machine rotation (αR). The minimum distance from the car frame rope-ends to the stator machine axis is named h_f. In a similar way, the minimum in-service distance from the counterweight rope ends to the stator machine axis is named h_w.

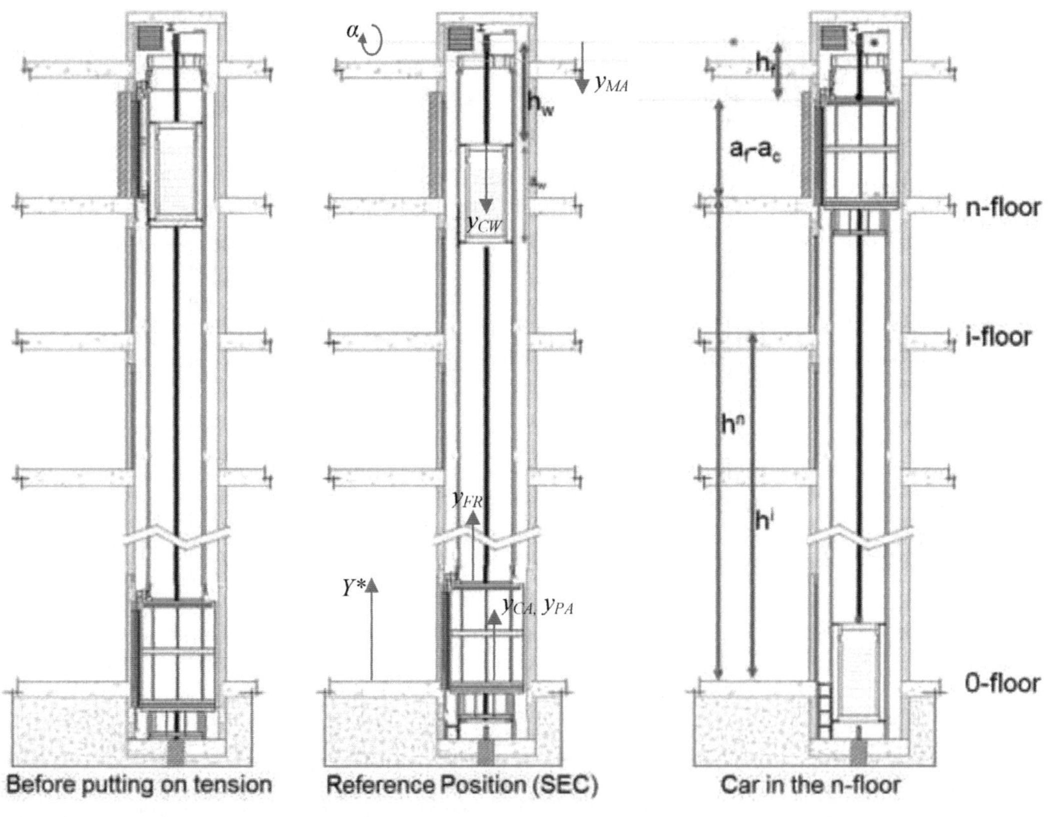

Figure 1. Installation parameters of the lift system

The center diagram in figure 1 shows the reference configuration of the elevator system (*SEC*) which is made coincident with the typical configuration after putting on tension the elevator in the installation process and rotate the drive, at a very low speed, to make the car floor level coincident with the *0*-floor level of the building.

From figure 1 is also possible to derive the steady-state configuration of the stopped lift with the machine brakes pressed at any car floor height y_{CA}. Then, the static deformations can be easily computed from the theory of elasticity, the sling properties (number of ropes n_R, net cross-sectional area of one rope A, rope density ρ, and instantaneous lengths l_f and l_w and equivalent stiffness k_{FR} and k_{CW} of both, frame rope-piece and counterweight rope-piece, respectively) the stiffness of the machine to vertical displacement k_{MA} and rotation with the brakes pressed k_{BR} and the car silent blocks stiffness k_{CA}. See table 1 in which m_{CW}, m_{FR}, m_{CA}, m_{PA} and m_{MA} are the masses of the counterweight, car frame, car, passengers/loads and machine, respectively. g is the gravity constant.

Table 1. Steady-state deformations of lift components under full car pressed brakes.

Component	Global position *[Y*]*	Initial Elongation/Rope lengths
V. disp.Machine	$h+h_f+a_f-a_c$	$y_{0MA}=\left(m_{MA}+m_{FR}+m_{CA}+m_{PA}+m_{CW}+n_R\rho A\left(h+h_f+h_c\right)\right)g/k_{MA}$
Counterweight	$y_{CA}+h+h_f+a_f-a_c-h_w$	$y_{0CW}=y_{0MA}+\left(m_{CW}+n_R\rho Al_w/2\right)g/k_{CW}+\alpha_0 R$
Car frame	$y_{CA}+a_f-a_c$	$y_{0FR}=-y_{0MA}-\left(m_{FR}+m_{CA}+m_{PA}+n_R\rho Al_f/2\right)g/k_{FR}+\alpha_0 R$
Car	y_{CA}	$y_{0CA}=y_{0FR}-\left(m_{CA}+m_{PA}\right)g/k_{CA}$
Pasengers/Loads	y_{CA}	$y_{0PA}=y_{0CA}-m_{PA}/k_{PA}$
Machine Rot.	--	$\alpha_0 R=\left(m_{CW}+n_R\rho A\left(l_w-l_f\right)-m_{FR}-m_{CA}-m_{PA}\right)g/k_{BR}$
Car frame rope	$l_f=h+h_f-y_{CA}$	$l'_f=l_f+\left(m_{FR}+m_{CA}+m_{PA}+n_R\rho Al_f/2\right)g/k_{FR}$
Counterw. rope	$l_w=h_c+y_{CA}$	$l'_w=l_w+\left(m_{CW}+n_R\rho Al_w/2\right)g/k_{CW}$

The derived mechanical model with 6 degrees of freedom [1] is shown in figure 2. The initial conditions of the system are made coincident with the equilibrium configuration of all the components during a stop after putting on tension the lift, thus with the brakes pressed, no gap between the car floor and the level of the landing floor and a critical time interval passed for assuming that the damping of the system makes the speed of all the components to be zero. Formulas in table 1 are based on the assumption that the machine brakes of a stopped lift behave like an elastic spring of constant k_{BR}. It is considered that the brakes take some time to grip and release, then the spring/damper constants are linearly variant with time from full intensity to zero during transition from stopped lift to moving lift and vice versa. Alternatively, it is possible to model the brake action using a brake torque T_{BR} which is applied to the traction pulley according to controlling parameters (On/off switch, rotation direction) and the active torque T_{ACT} exerted by the gravity forces and motor drive on the traction pulley. Then, the initial rotation $\alpha_0 R$ shown in table 1 is set to *0*, and constants k_{BR} and c_{BR} are inexistents in the model.

A standard trip consists basically of three stages. In the first stage the passengers/loads get on the lift car. This makes the system to vibrate together with the passengers/load and after a certain time lapse Δt_1 the car-frame sling will elongate

$$\Delta l_{1f}=\frac{m_{PA}g}{n_R EA}l_f \tag{1}$$

where n_R is the number of ropes, EA is the longitudinal rigidity of one rope. The gap between car-floor and landing will be:

$$y_{CA}^{1}(\Delta t_1) = y_{CA}^{1}(0) - \Delta l_{1f} - \frac{m_{PA}g}{k_{CA}} - \frac{m_{PA}g}{k_{MA}} \tag{2}$$

where k_{CA} and k_{MA} are the stiffness of the car and machine silent-blocks, respectively.

The second stage starts at time lapse t_i when the button of the desired floor is pressed. Then, a series of electric and electronic components called the controller are responsible to release the brakes and compute the necessary torque T for the car to follow the setpoint speed for the desired trip. This stage finishes at time lapse t_f once the setpoint car speed gets zero, the brakes are pressed and motor switch is set to off.

In the third stage the passengers/loads get off the car and the system vibrates during some time Δt_3. The gap between car-floor and landing will be:

$$y_{CA}^{3}(t_f + \Delta t_3) = y_{CA}^{3}(t_f) + \Delta l_{1f} + \frac{m_{PA}g}{k_{CA}} + \frac{m_{PA}g}{k_{MA}} \tag{3}$$

The vibrations and response of the system in the three stages is obtained by the integration of the governing equations following the step by step process indicated in the next paragraph.

The traction forces on the rope have to be transmitted through friction grip between the traction sheave and the rope, resulting inevitably in slip. Following Berner [2] the slip is considered vanishing,

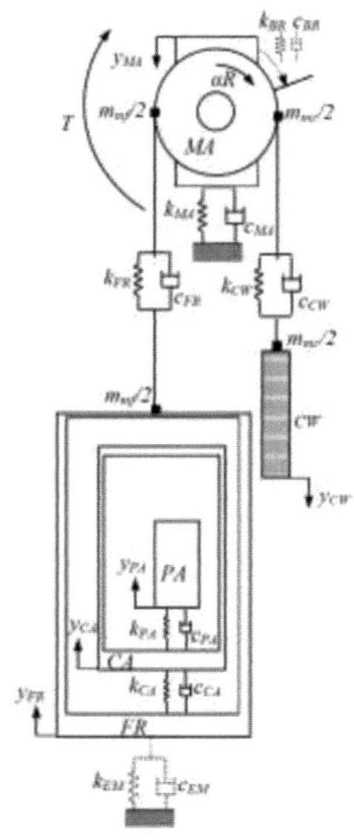

Figure 2. The 6 d.o.f mechanical model of the proposed elevator car system [1].

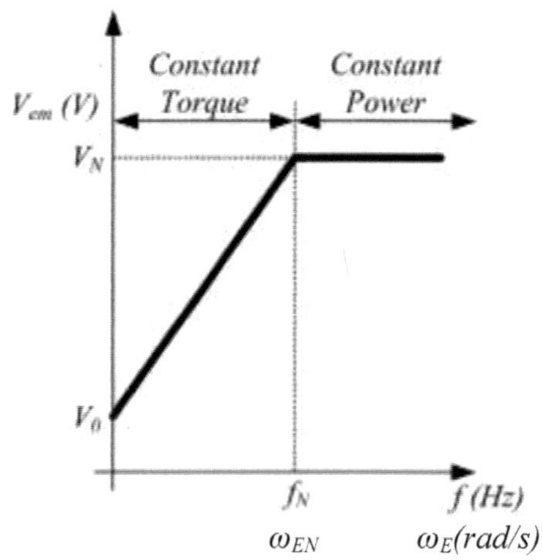

Figure 3. Scalar or V/f law control.

that's consistent with rope force ratios close to 1 and no gilding slip occurs.

3. Step by step process

The system of equations that govern the elevator system [1] are:

$$[M][\ddot{y}]+[C][\dot{y}]+[K][y]=[F] \tag{4}$$

where $[M]$ is the mass matrix, $[C]$ is the damping matrix, $[K]$ is the stiffness matrix, $[F]$ is the time dependent vector of applied forces and $[y]$ is the displacement vector:

$$[y]=\begin{bmatrix}\alpha R & y_{FR} & y_{CA} & y_{PA} & y_{CW} & y_{MA}\end{bmatrix}^T \tag{5}$$

The system of equations can be solved using the iterative method of Runge-Kutta[3]. This method can be applied easily in MATLAB after some preparation. First, we obtain the acceleration vector multiplying the members of the system of equations (4) by the inverse of the matrix $[M]$:

$$[\ddot{y}]=[M]^{-1}\{[F]-[C][\dot{y}]-[K][y]\} \tag{6}$$

We define a vector of unknowns $[Y(t)]$ including the displacement vector $[y]$ and its derivate:

$$[Y(t)]=\begin{bmatrix}[y]\\ [\dot{y}]\end{bmatrix} \tag{7}$$

Then the derivate of $[Y(t)]$ can be expressed in the form:

$$\left[\dot{Y}(t)\right]=\begin{bmatrix}[\dot{y}]\\ [\ddot{y}]\end{bmatrix}=\begin{bmatrix}[\dot{y}]\\ [M]^{-1}\{[F]-[C][\dot{y}]-[K][y]\}\end{bmatrix} \tag{8}$$

whose terms can be rearranged:

$$\left[\dot{Y}(t)\right]=\begin{bmatrix}[0]_{6x6} & [I]_{6x6}\\ -[M]^{-1}[K] & -[M]^{-1}[C]\end{bmatrix}\begin{bmatrix}[y]\\ [\dot{y}]\end{bmatrix}+\begin{bmatrix}[0]\\ [M]^{-1}[F]\end{bmatrix} \tag{9}$$

If we call $[A]$ to the first square matrix in (9), which is nearly a time-constant matrix, and $[B(t)]$ to the vector added to the right, the expression (9) gives:

$$\left[\dot{Y}(t)\right]=[A][Y(t)]+[B(t)] \tag{10}$$

The vector $[B(t)]$, is a function of the excitation forces $[F(t)]$ which has the form [1]:

$$[F]=\begin{bmatrix}\frac{T}{R}+\left(n_R\rho Al_w/2-n_R\rho Al_f/2\right)g\\ -\left(m_{FR}+n_R\rho Al_f/2\right)g\\ -m_{CA}g\\ -m_{PA}g\\ \left(m_{CW}+n_R\rho Al_w/2\right)g\\ \left(m_{MA}+n_R\rho Al_f/2+n_R\rho Al_w/2\right)g\end{bmatrix} \tag{11}$$

where T represents the applied torque (which is the sum of that produced by the electric machine T_{EL} and that of the brakes T_{BR}) and will depend on the setpoint speed v_{ref} according to the control law established by the electronic controller and on the electrical machine parameters, and R is the traction sheave radius. It is clear that the gravity torque and equivalent gravity forces on the car-frame and counterweight are not time-constant because the assumed rotation with no-slip of the traction sheave makes the rope mass to be transferred from one suspension side to the other.

Once the lift controller fixes the torque T to be exerted at a certain step i ($t=t_i$) according to next paragraph, the set of governing equations (10) can be solved using the MATLAB function *ode45*. Then the position and velocities of all the components (7) are determined in the next step ($t=t_i+\Delta t$) where Δt is the time increment of the integration process.

4. The lift controller

As a first simulation step, and ideal controller which computes the motor torque according to kinetic momentum equation of the rotor machine drive as a function of the preset acceleration of the setpoint speed v_{ref} profile.

A second simulation step consists on the modelling of an electronic converter which is assumed to have a basic scalar (or V/f law) control. Its main function is to reconstruct a three-phase sinusoidal voltage controlling the RMS and frequency of the sinusoidal waveforms according to figure 3. This control operates the electrical motor in two zones, one named "Constant torque" and the other named "Constant power" where the control tries to maintain the torque (or control) constant in the electric machine. It has also to be highlighted that the applied voltage at zero speed is not zero, but a minimum value, V_0, in order to maintain the torque at low speeds [4-8].

The electronic control analyses the starting and ending point of the vertical trip, and establish the car speed profile for the setpoint speed v_{ref}, reacting to limit switches and actuating to press or release the mechanical brakes. Once the speed profile for the trip $v_{ref}= v_{ref}(t)$ is established, the electronic control generates the sinusoidal voltage to be applied to the electrical motor at time t_i, according to next equations:

$$\omega_E = n_p \cdot v_{ref} \cdot k_L \cdot k_R \tag{12}$$

$$V_{em} = V_0 + (V_N - V_0)\frac{|\omega_E|}{\omega_{EN}} \tag{13}$$

where ω_E is the angular speed of the voltage applied to the electric motor (adopted in absolute value as far as negative frequencies require positive voltages, and the interchange of the supply three-phase sequence), $n_P \cdot k_L \cdot k_R$ is the factor to translate the vertical speed of the car v_{ref} into electrical motor angular speed ω_E taking into account the number of pair of the asynchronous or induction electric motor n_p, the index ratio of the machine gear box k_R and the conversion factor from the vertical speed of the car and the mechanical angular speed of the traction sheave k_L ($k_L=n/$R for an $n{:}1$ roping elevator), V_{em} is the *RMS* value of the voltage applied to the electric motor; V_N , f_N and ω_{EN} are the nominal *RMS* voltage, frequency and angular speed of the electric motor, and V_0 is the minimum voltage applied by the electronic converter at zero speed. The electric motor is modelled as an instantaneous torque source T_{EL}, with next expression:

$$T_{EL} = k_R \frac{m_1 \frac{R_2'}{s}}{\omega_E + \omega_\varepsilon} \frac{V_{em}^2}{\left(\left(R_1 + \frac{R_2'}{s}\right)^2 + \left(L_{cc} \cdot \omega_E\right)^2\right)} \tag{14}$$

where m_1 is the number of phases (3 in most cases), R_1 is the stator resistance, R'_2 is the rotor resistance passed to stator, L_{cc} is the electric machine series inductance and s is the electric machine slip given by:

$$s = \frac{\omega_E - n_p \cdot v_{mec} \cdot k_L \cdot k_R}{\omega_E} \tag{15}$$

where v_{mec} is the mechanical speed of the vertical movement. In (14) a value of ω_ε=0.001rad/s is included to prevent dividing by zero for zero angular speed ω_E case. That is different from the set point speed v_{ref}, as the induction machine operates with the previously indicated slip, which is dependent on the torque the machine is producing. The mechanical speed v_{mec} is frequently made coincident with the tangent speed of the traction sheave $\dot{\alpha}R$ which can be logged out easily by means of an encoder attached to the machine stator in engineering practice. We take $\dot{\alpha}R$ from the seventh component of the solution vector $[Y(t_i)]$ at time t_i. Another option is to make the mechanical speed coincident with that of the car speed $\dot{y}_{CA}(t_i)$.

5. Computing a trip

The step by step process starts in the stopped configuration. The spring constant k_{BR} is fixed to its nominal value typically $k_{BR}=10^8 N/m$ in the option of modelling the brake actuation by a spring/damper constant. That can be estimated by $T_{BR}/(\varphi R^2)$ where T_{BR} is the torque exerted by the brakes on the motor and φ is the angle during the braking action). Then, preset time variant passengers mass $m_{PA}(t)$ and spring constant $k_{BR}(t)$ are introduced to simulate the getting on process. In the other option, k_{BR} is set to 0. Then the brake torque T_{BR} is computed from the active torque T_{ACT} and the instant maximum braking torque at every step.

"Get on" and "Get off" routines are used before and after any standard trip. Initial conditions previous to getting on or getting off the car can be established by table 1 for the empty car ($m_{PA}=0$) and stopped system (brakes pressed).

The simulated standard trip stage is the main routine in the program. The trip is simulated from *i-th* floor to *j-th* floor of an installation using MATLAB [9]. The program follows the step by step process previously described then αR, y_{CA}, y_{MA}, y_{FR}, y_{PA} , y_{CW} and its first, second and third derivates are estimated during the time interval $[t_i;t_f]$ in time increments Δt. A general integration time lapse Δt of 0.005s is adopted.

Now we are ready to compare the performance of the various roping systems at the functional differences in the elevator installation.

6. Results

Two typical load/unload processes were analyzed. In the passengers' load process the passenger mass m_{PA} is varying from 0 to its nominal value at the certain rate (typically 0.8-1 passengers per second [10]). Goods' loading process is considered too. Handling trucks or freight cars loading is exerted in

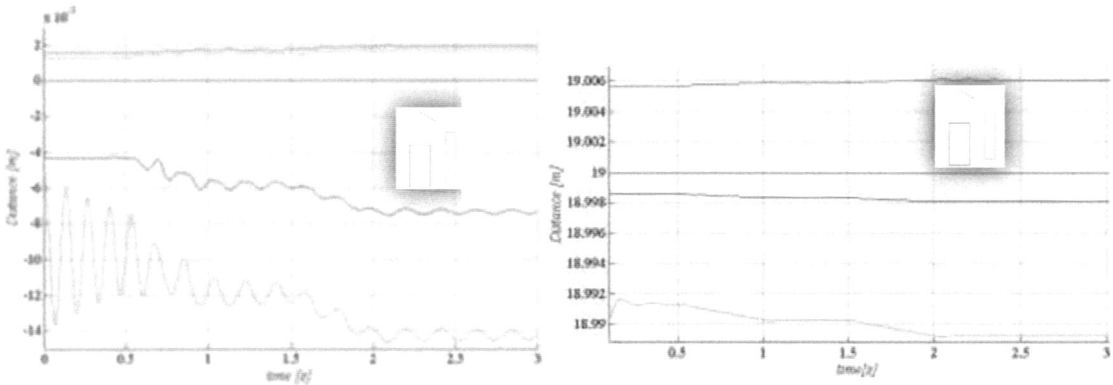

Figure 4. Vertical displacements during a residential elevator getting on process. (right) 8 passengers get on floor 6 (19m height) at 8 passengers per second rate: [0.5s-1s] 4 passengers are getting on [1s-1.5s] stay, [1.5s-2s] Additional 4 passengers getting on. (left) 450 kg palette truck get on 0-floor (0m height): [0.5-1s] front wheels are getting on, [1.5s-2s] rear wheels are getting on.

two steps: First, the front wheels are lifted up and then, after some time interval, the rear wheels are lifted up too.

Then, the plot of the instantaneous loaded mass $m_{PA}(t)$ vs time t adopts a bilinear form interlaced with a constant mass period. Impacts during getting on/off the car is simulated using a passenger mass m_{PA} weighted by an equivalent dynamic/static impact coefficient during a short time interval.

During the load/unload stages all the lift components influences its mechanical response, the car and machine silent-blocks are strained and all the components vibrate vertically to some extent. The change in gap between landing floor and car-floor can be easily computed from the displacement plots comparing the initial ($t=0$) and end values ($t=\Delta t_1$) of the car displacement y_{CA}.

The step by step process indicated in previous paragraph starts once the stopped configuration, the load/unload type and the time variant passengers mass $m_{PA}(t)$ are adopted. A general integration time

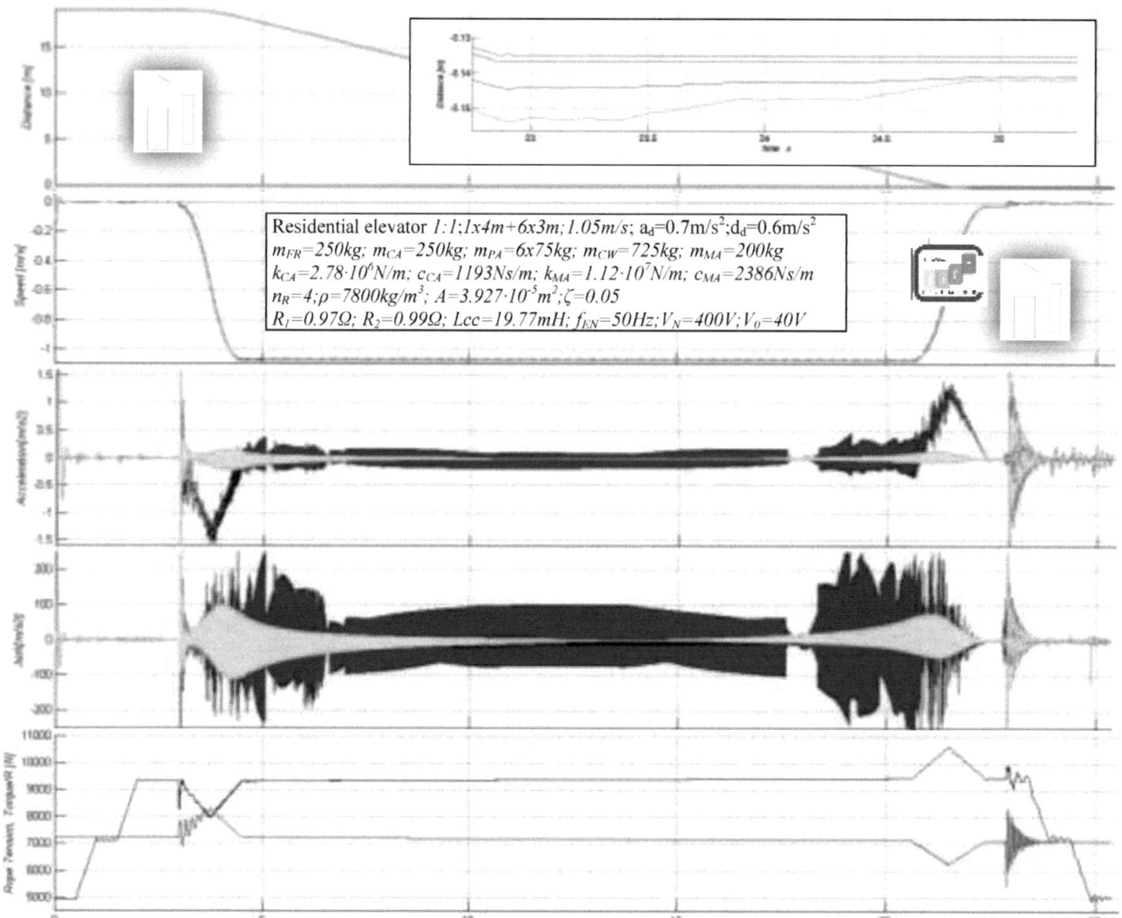

Figure 5. Full passengers' trip of an 8 passengers residential elevator at a rated speed of 1.05m/s from 6th to 0-floor inspired in the performance of Ziehl-Abegg's asynchronous machine drive (VFD132.M-4) and using scalar control routine. (Extra key colors, magenta: prescribed speed)

lapse Δt of 0.005s is initially set. After a period of computing time, the display will show the response of the system (figure 4), where we can see the system oscillations to reach balance or the steady state response during the load process.

The results are shown in the computer display (Figure 5), where the position, velocity, acceleration and jerk of each component and the slings tension during the full trip are plotted. Using MATLAB zoom capabilities of the plotted figures a more detailed description of velocity, for example, can be observed (figure 6). The frequency spectrum of all mechanical variables and components during a selectable time gap can be plotted (figure 7). It is observed a different frequency spectrum whether data includes the overall trip or only those from the constant speed portion.

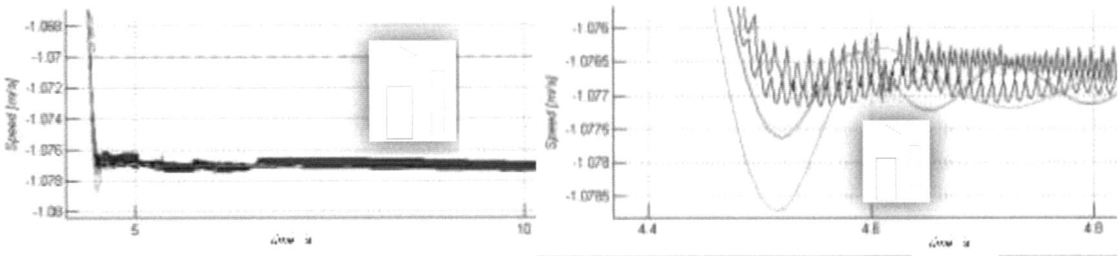

Figure 6. Left: Zoom of the speed time history in figure 5 showing descrepancies to preset speed plot (-- magenta) with increasing time. Right: Extra zoom of the plot in the left showing detailed component speeds.

Qualitative validation of the simulations for a number of characteristics scenarios has been made. A series of elementary validation tests were done by assigning an infinite stiffness value to all the stiffness parameters of the system except the ropes. These elements also had been assigned a minimum damping coefficient. Attending to the results, the program is fully validated with an overall error of about 0.3%.

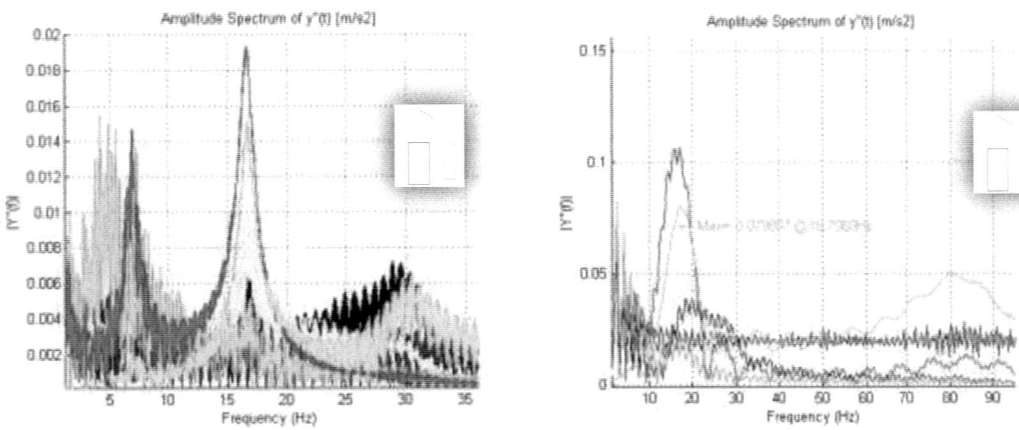

Figure 7. Acceleration frequency spectrum of (left) the overall trip shown in figure 5 (right) constant speed travel [5-20s] portion.

7. Conclusions

A method for solving interactively the equations of motion of any specified lift system by means of a computer has been derived and implemented in the MATLAB platform. A number of dynamics phenomena concerning car position, acceleration, speed or jerk that occurs during the normal functioning in a standard trip and other related exceptional circumstances like impacts provided by guide misalignment or uncontrolled car movements can be reproduced with enough accuracy.

Scalar control and static model have been analysed for electronic controller an electric machine. In future work, more precise controls can be analysed, as field oriented control (FOC), and more detailed model including the dynamics of the electric machine can be applied to analyse the behaviour of the whole system deeply. Moreover, by using FOC control, advanced control techniques, as active damping of mechanic vibration, can be modelled to study how these techniques can mitigate undesired effect in the whole system and how a more precise and smooth control during a standard trip can be achieved.

In future work more detailed and demanding validation tests can be performed to make test tower experiments compatible to simulated results.

Acknowledgements
Authors would like to thank Mr. Javier García Garralón from Ziehl-Abegg Spain for his help and colaboration in the compilation of machine drive and brakes data and the modelling of electric parameters in real elevator machine drives.

References

[1] Sicilia J R, Alonso F J and Herrera I 2015 Mechanical Discrete Simulator of the electro-mechanical lift with n:1 roping *Symposium on the Mechanics of Slender Structures 21-22 September 2015 Northampton UK.*
[2] Berner O 2011 Diss. *Institut Für Fördertechnik und Logistik* University of Sttutgart June 2011
[3] Rao SS Mechanical Vibrations 2005 (Singapore: Prentice Hall)
[4] Waheedabeevi M, Sukeshkumar A and Nair N S 2012 New online loss-minimization-based control of scalar and vector-controlled induction motor drives *Power Electronics, Drives and Energy Systems (PEDES), 2012 IEEE International Conference on* 1,7 16-19 Dec 2012
[5] Pati S, Mohanty S and Patnaik S 2014 Improvement of transient and steady state performance of a scalar controlled induction motor using sliding mode controller *Circuit, Power and Computing Technologies (ICCPCT), 2014 International Conference on* 20-21 220,225 March 2014
[6] Syed F U, Kuang M L and Ying H 2009 Active Damping Wheel-Torque Control System to Reduce Driveline Oscillations in a Power-Split Hybrid Electric Vehicle *Vehicular Technology IEEE Transactions on* 58 9 4769-4785 Nov. 2009
[7] Farid N, Sebti B, Mebarka K and Tayeb B 2007 Performance analysis of field-oriented control and direct torque control for sensorless induction motor drives *Control & Automation 2007 MED '07 Mediterranean Conference on* 1 6 27-29 June 2007
[8] Sira-Ramirez H, Gonzalez-Montanez F et al 2013 A Robust Linear Field-Oriented Voltage Control for the Induction Motor: Experimental Results *Industrial Electronics IEEE Transactions on* 60 8 3025-3033 Aug. 2013 doi: 10.1109/TIE.2012.2201430
[9] Herrera I 2010 Mechanical Modelling and Simulation Lift Systems: the Influence of the Dynamic Behaviour of Slender Structural Components *Moss 2010 Symposium on the mechanics of slender structures* ISBN.978.84.608.1100.8
[10] Strakosh G R and Caporale R S 2010 The Vertical Transportation Handbook *John Wiley and Sons* 2010 Page 75.

Dynamic response of mechanical systems to impulse process stochastic excitations: Markov approach

R Iwankiewicz

Hamburg University of Technology, Eissendorfer Strasse 42, 21073 Hamburg, Germany and University of Zielona Gora, ul. Prof. Szafrana 1, 65-516 Zielona Gora, Poland

E-mail: iwankiewicz@tu-harburg.de

Abstract. Methods for determination of the response of mechanical dynamic systems to Poisson and non-Poisson impulse process stochastic excitations are presented. Stochastic differential and integro-differential equations of motion are introduced. For systems driven by Poisson impulse process the tools of the theory of non-diffusive Markov processes are used. These are: the generalized Itô's differential rule which allows to derive the differential equations for response moments and the forward integro-differential Chapman-Kolmogorov equation from which the equation governing the probability density of the response is obtained. The relation of Poisson impulse process problems to the theory of diffusive Markov processes is given. For systems driven by a class of non-Poisson (Erlang renewal) impulse processes an exact conversion of the original non-Markov problem into a Markov one is based on the appended Markov chain corresponding to the introduced auxiliary pure jump stochastic process. The derivation of the set of integro-differential equations for response probability density and also a moment equations technique are based on the forward integro-differential Chapman-Kolmogorov equation. An illustrating numerical example is also included.

1. Introduction

In some problems in structural dynamics the loads may be adequately idealized as discontinuous stochastic processes: trains of loading pulses, or impulses, occurring at random times and characterized by random magnitudes. Such models are pertinent e.g. to: behaviour of a vehicle travelling over a very rough road (impacts or shocks, due to sudden humps or holes) [1,2], moving loads on a bridge due to highway traffic [3,4], dynamic loading due to wind gusts [5-7]. It is known that if the excitation is a Poisson impulse process, the state vector of the dynamic system is non-diffusive (Poisson-driven)a Markov process. However, if the excitation is a non-Poisson (e.g. renewal) impulse process, the state vector of the dynamic system is not a Markov process. Conversion of the original non-Markov pulse problem into a Markov one is in some cases possible owing to the introduction of auxiliary state variables in form of pure-jump stochastic processes governed by Poisson-driven stochastic differential equations. Then the original state vector augmented by those auxiliary variables becomes a non-diffusive Markov process and differential equations for moments may be derived [8-12]. At the same time, the explicitly introduced, Poisson-driven, pure-jump stochastic processes are characterized by a chain of Markov states. Consequently the original state variables and the states of the auxiliary pure-jump stochastic process are jointly Markovian and the problem is characterized by the set of mixed-type, joint probability density - discrete distribution functions governed by forward integro-differential Chapman-Kolmogorov equations. From the forward equation, with the aid of suitably defined jump probability intensity functions, the explicit integro-differential equations for the response probability density function are derived [13]. The response statistical moments are defined as integrals with respect to the mixed-type, joint probability density - discrete distribution function. Based on the forward integro-differential Chapman-Kolmogorov equation the differential equations for moments are obtained [14]. An illustrating numerical example is also included.

2. State-space formulation of the problem: stochastic differential equations of motion

When the pulse duration is much shorter that the fundamental natural period of the system the train of short-duration pulses is idealized as an impulse process

$$F(t) = \sum_{i=1}^{\mathcal{N}(t)} P_i \delta(t - t_i), \tag{1}$$

where $\mathcal{N}(t)$ is a stochastic point (random counting) process (e.g. a Poisson or a renewal process.), $t_i \in [0, t)$, i.e. excluding the event (impulse) that may occur at t. The impulse magnitudes P_i are independent random variables, identically distributed as a random variable P and statistically independent of the random times t_i, or of the counting process $\mathcal{N}(t)$. The usual differential

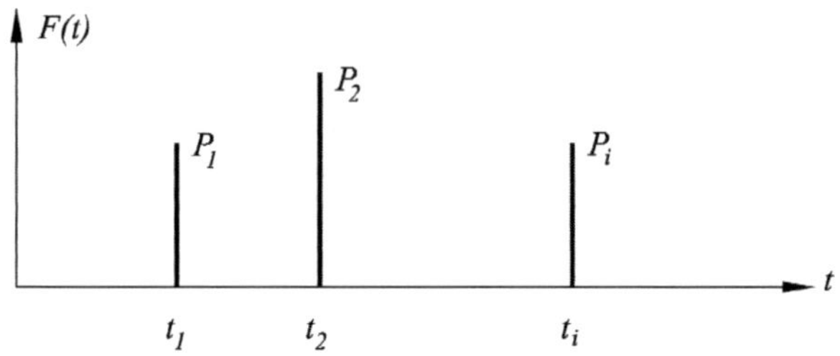

Figure 1. Stochastic impulse process

equation of motion of a linear SDOF system is

$$\ddot{Y}(t) + 2\zeta\omega\dot{Y}(t) + \omega^2 Y(t) = \sum_{i=1}^{\mathcal{N}(t)} P_i \delta(t - t_i) \tag{2}$$

From the impulse-momentum principle it follows that at the impulses occurrence times t_i the velocity response process $\dot{Y}(t)$ changes by jumps, it is piece-wise continuous (continuous-jump). Consequently the displacement response process $Y(t)$ is continuous but it is only piece-wise continuously differentiable. As the jump change in the velocity may occur at any point, the usual rules of calculus do not apply. The usual notation of the differential equation of motion is **not mathematically meaningful**. The stochastic counterparts of the usual differential equations are the **stochastic differential equations**

$$\begin{aligned} dY(t) &= \dot{Y}(t)dt, \\ d\dot{Y}(t) &= -2\zeta\omega\dot{Y}(t)dt - \omega^2 Y(t)dt + P(t)d\mathcal{N}(t), \end{aligned} \tag{3}$$

where $P(t)$ is the magnitude of the impulse which occurs in the time interval $[t, t + dt)$.

The impulse process excitation may be also parametric, i.e. multiplicative to the displacement or to the velocity response process.

Consider the stochastic impulse process $F(t)$ acting as an axial force on a beam-column (Figure 2). The governing equation of motion is

$$EI\frac{\partial^4 w(x,t)}{\partial x^4}+c\frac{\partial w(x,t)}{\partial t}+\mu\frac{\partial^2 w(x,t)}{\partial t^2}+F(t)\frac{\partial^2 w(x,t)}{\partial x^2}=0,\quad F(t)=\sum_{i=1}^{\mathcal{N}(t)}P_i\delta(t-t_i) \tag{4}$$

If the normal mode approach is used

$$w(x,t)=\sum_{j=1}^{n}Y_j(t)\varphi_j(x), \tag{5}$$

where $\varphi_j(x)$ are the normal modes then the set of usual differential equation of motion is

$$\ddot{Y}_j(t)+2\zeta_j\omega_j\dot{Y}_j(t)+\omega_j^2Y_j(t)+\beta_jY_j(t)\sum_{i=1}^{\mathcal{N}(t)}P_i\delta(t-t_i)=0,\quad j=1,2,3,...n, \tag{6}$$

where $\beta_j=\dfrac{\int_o^\ell \varphi_j''(x)\varphi_j(x)dx}{\mu\int_o^\ell \varphi_j^2(x)dx}$ and $(...)''=\dfrac{d^2}{dx^2}(...)$.

The stochastic differential equations of motion are obtained as

$$\begin{aligned} dY_j(t) &= \dot{Y}_j(t)dt, \\ d\dot{Y}_j(t) &= -2\zeta_j\omega_j\dot{Y}_j(t)dt-\omega_j^2Y_j(t)dt-\beta_jY_j(t)P(t)d\mathcal{N}(t),\quad j=1,2,3,...n. \end{aligned} \tag{7}$$

hence the impulse process excitation term is multiplicative to the displacement response process.

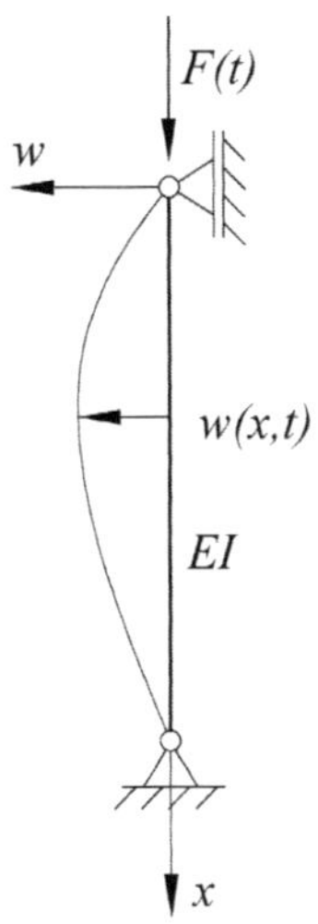

Figure 2. Stochastic impulse process $F(t)$ acting as an axial force on a beam-column.

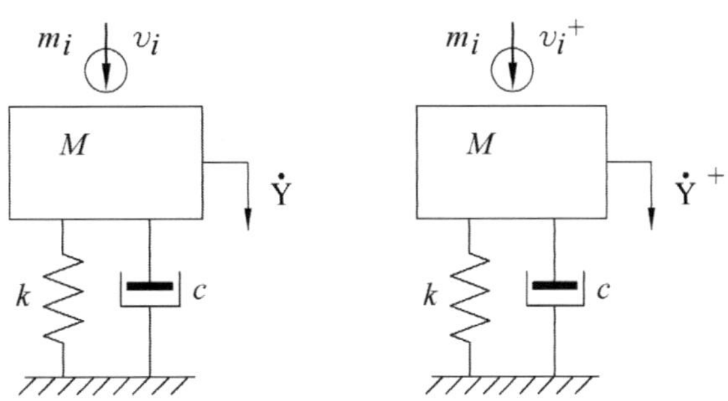

Figure 3. Linear oscillator of mass M subjected to a random train of impacts with particles of random masses m_i, arriving at random times t_i with random velocities v_i.

Consider now a linear oscillator of mass M is subjected to a random train of impacts with particles of random masses m_i, arriving at random times t_i with random velocities v_i [15] as shown in the Figure 3. Before the impact the oscillator has the velocity $\dot{Y}$. Random times t_i are assumed to be independent of m_i and v_i. After the impact the velocities of the oscillator and of the particle are $\dot{Y}^+$ and v_i^+, respectively.

The law of conservation of linear momentum states that

$$M\dot{Y} + m_i v_i = M\dot{Y}^+ + m_i v_i^+ \tag{8}$$

At the time of impact the jump change in the oscillator velocity equals

$$\dot{Y}^+ - \dot{Y} = \frac{m_i(1+K)}{M+m_i}(v_i - \dot{Y}). \tag{9}$$

With due account for the impulse of restoring and damping forces during the time interval dt the stochastic equations of motion are

$$\begin{aligned} dY &= \dot{Y}dt, \\ d\dot{Y} &= (-2\zeta\omega\dot{Y} - \omega^2 Y)dt + (V(t) - \dot{Y})P(t)d\mathcal{N}(t), \end{aligned} \tag{10}$$

where the random variable equals $P(t) = \dfrac{m(t)(1+K)}{M+m(t)}$ and $m(t)$ and $V(t)$ denote, respectively, the mass and the velocity of the particle which impacts in the time interval $[t, t+dt)$ and K is the coefficient of restitution. Hence the impulse process excitation terms are additive and multiplicative to the velocity response process.

In general the state vector of the system, $\mathbf{Y}(t)$, consisting of the generalized coordinates and velocities, is governed by the set of stochastic differential equations of motion

$$d\mathbf{Y}(t) = \mathbf{c}(\mathbf{Y}(t), t)dt + \mathbf{b}(\mathbf{Y}(t), t))P(t)d\mathcal{N}(t), \quad \mathbf{Y}(0) = \mathbf{y}_0, \tag{11}$$

3. Non-diffusive Markov problem

3.1. Stochastic integro-differential equation of motion

If $\mathcal{N}(t)$ is a Poisson counting process $N(t)$ (hence the increments $dN(t)$ are independent), $N(t)$ and P_i are independent, P_i are mutually independent and both $N(t)$ and P_i are independent of the initial conditions $\mathbf{y}_0$, then the state vector $\mathbf{Y}(t)$ is a Markov process. It is then a **non-diffusive (Poisson-driven) Markov process** governed by the set of stochastic integro-differential equations [16, 17]

$$d\mathbf{Y}(t) = \mathbf{c}(\mathbf{Y}(t), t)dt + \mathbf{b}(\mathbf{Y}(t), t)) \int_{\mathcal{P}} \mathrm{p}\Pi(dt, dp), \quad \mathbf{Y}(0) = \mathbf{y}_0, \tag{12}$$

where $\mathbf{c}(\mathbf{Y}(t), t)$ is the drift vector, $\mathbf{b}(\mathbf{Y}(t), t))$ is an analogue of the diffusion vector, $\Pi(dt, dp)$ is the Poisson random measure, interpreted as the random number of points (impulses) in the interval $[t, t+dt)$ with the values of the random impulse magnitude P in $(p, p+dp)$ and $\mathcal{P}$ is the sample space of the random impulse magnitude P.

3.2. Generalized Itô's differential rule

For a function $V[t, \mathbf{Y}(t)]$, bounded for t and $\mathbf{Y}(t)$ finite, and once continuously differentiable with respect to all its arguments, the following **generalized Itô's differential rule** is valid for a Poisson-driven Markov process [16, 17]

$$
\begin{aligned}
dV(t,\mathbf{Y}(t)) &= \frac{\partial V(t,\mathbf{Y}(t))}{\partial t}dt + \sum_{i=1}^{n}\frac{\partial V(t,\mathbf{Y}(t))}{\partial Y_i}c_i\left(\mathbf{Y}(t),t\right)dt \\
&+ \int_{\mathcal{P}}\left[V\left[t,\mathbf{Y}(t)+\mathbf{b}\big(\mathbf{Y}(t),t\big)p\right] - V\left[t,\mathbf{Y}(t)\right]\right]\Pi(dt,dp)
\end{aligned}
\tag{13}
$$

The generating equation for moments is obtained as

$$
\begin{aligned}
\frac{d}{dt}E\left[V(\mathbf{Y}(t))\right] &= \sum_{i=1}^{n}E\left[\frac{\partial V(\mathbf{Y}(t))}{\partial Y_i}c_i\left(\mathbf{Y}(t),t\right)\right] \\
&+\nu\int_{\mathcal{P}}E\Big[V\left[\mathbf{Y}(t)+\mathbf{b}\big(\mathbf{Y}(t)\big)p\right]\Big]f_P(p)dp - \nu E\Big[V\left[\mathbf{Y}(t)\right]\Big],
\end{aligned}
\tag{14}
$$

where $f_P(p)$ is the probability density function of the random impulse magnitude P. Explicit equations for moments are obtained for $V[\mathbf{Y}(t)] = Y_i(t)Y_j(t)$, $V[\mathbf{Y}(t)] = Y_i(t)Y_j(t)Y_k(t)$, $V[\mathbf{Y}(t)] = Y_i(t)Y_j(t)Y_k(t)Y_l(t)$, etc.

The equations for the mean values, the second-, third- and fourth-order joint central moments of the response, are obtained as

$$
\dot{\mu}_i(t) = E\left[c_i(\mathbf{Y}(t),t)\right] + \nu(t)E[P]E\left[b_i(\mathbf{Y}(t),t)\right] \tag{15}
$$

$$
\begin{aligned}
\dot{\kappa}_{ij}(t) &= 2\Big\{E\Big[Y_i^0\Big(c_j^0(\mathbf{Y}^0(t),t)+\nu(t)b_j(\mathbf{Y}(t),t)P\Big)\Big]\Big\}_s \\
&+\nu(t)E[P^2]E\left[b_i(\mathbf{Y}(t),t)b_j(\mathbf{Y}(t),t)\right]
\end{aligned}
\tag{16}
$$

$$
\begin{aligned}
\dot{\kappa}_{ijk}(t) &= 3\Big\{E\Big[Y_i^0Y_j^0\Big(c_k^0(\mathbf{Y}^0(t),t)+\nu(t)b_k(\mathbf{Y}(t),t)P\Big)\Big]\Big\}_s \\
&+3\nu(t)E[P^2]\Big\{E\Big[Y_i^0b_j(\mathbf{Y}(t),t)b_k(\mathbf{Y}(t),t)\Big]\Big\}_s \\
&+\nu(t)E[P^3]E\left[b_i(\mathbf{Y}(t),t)b_j(\mathbf{Y}(t),t)b_k(\mathbf{Y}(t),t)\right]
\end{aligned}
\tag{17}
$$

$$
\begin{aligned}
\dot{\kappa}_{ijkl}(t) &= 4\Big\{E\Big[Y_i^0Y_j^0Y_k^0\Big(c_l^0(\mathbf{Y}^0(t),t)+\nu(t)b_l(\mathbf{Y}(t),t)P\Big)\Big]\Big\}_s \\
&+6\nu(t)E[P^2]\Big\{E\Big[Y_i^0Y_j^0b_k(\mathbf{Y}(t),t)b_l(\mathbf{Y}(t),t)\Big]\Big\}_s \\
&+4\nu(t)E[P^3]\Big\{E\Big[Y_i^0b_j(\mathbf{Y}(t),t)b_k(\mathbf{Y}(t),t)b_l(\mathbf{Y}(t),t)\Big]\Big\}_s \\
&+\nu(t)E[P^4]E\left[b_i(\mathbf{Y}(t),t)b_j(\mathbf{Y}(t),t)b_k(\mathbf{Y}(t),t)b_l(\mathbf{Y}(t),t)\right],
\end{aligned}
\tag{18}
$$

where $Y_i^0(t) = Y_i(t) - \mu_i(t)$ and $c_j^0(\mathbf{Y}^0(t),t) = c_j(\mathbf{Y}(t),t) - E\left[c_j(\mathbf{Y}(t),t)\right]$ denote the zero-mean (centralized) state variables and drift vector, and $\{...\}_s$ denotes the Stratonovich symmetrizing operation, e.g.

$$
\{Y_iY_jc_k\}_s = \frac{1}{3}(Y_iY_jc_k + Y_iY_kc_j + Y_jY_kc_i) \tag{19}
$$

Equations for response moments of a linear system form always a closed set and can be directly solved numerically.

However, equations for response moments of a non-linear system with polynomial non-linearity form an infinite hierarchy. Then special closure approximations (truncation procedures) must be used, which express higher-order moments (above the level of truncation) in terms of lower-order moments (up to the level of truncation). For other types of non-linearity the equations involve unknown expectations of non-linear functions of state variables and some tentative forms of the joint probability density must be used.

For example if the fifth- and sixth-order cumulants are neglected, the fifth- and sixth-order central moments are expressed in terms of lower-order moments as

$$\begin{aligned} \kappa_{ijklm}(t) &= 10\left\{\kappa_{ij}(t)\kappa_{klm}(t)\right\}_s \\ \kappa_{ijklmn}(t) &= 15\left\{\kappa_{ij}(t)\kappa_{klmn}(t)\right\}_s + 10\left\{\kappa_{ijk}(t)\kappa_{lmn}(t)\right\}_s - 30\left\{\kappa_{ij}(t)\kappa_{kl}(t)\kappa_{mn}(t)\right\}_s \end{aligned} \tag{20}$$

The above closure approximations are equivalent to assuming the tentative joint probability density function of the state variables in the form of the suitable Gram-Charlier expansion, which is the expansion in terms of generalized Hermite polynomials. The accuracy of the results obtained with the help of closure approximations depends on whether or not the truncated pertinent Gram-Charlier expansion is an adequate approximation to the actual density function.

Consider the dynamical system subjected to a random train of impulses, with zero initial conditions $\mathbf{Y}(0) = \mathbf{0}$. If in the time interval $[0, t)$ no impulse occurred, the system is still at rest at the time t.

If the train of impulses is driven by a homogeneous Poisson process, the probability P_0 of no impulse occurrence in the time interval $[0, t)$ is expressed as

$$P_0 = \Pr\{N(t) = 0\} = \exp(-\nu t). \tag{21}$$

The probability P_0 may be high, close to the unity, if the length t of the time interval is small, i.e. at the early transient stage, especially if also the mean arrival rate ν is small.

Joint probability density function of the state vector $\mathbf{Y}(t)$ can be represented in form of the sum of the continuous and discrete parts as (Figure 4) [18]

$$\begin{aligned} f_{\mathbf{Y}}(\mathbf{y}, t) &= f_{\mathbf{Y}}(\mathbf{y}, t \mid N(t) = 0)\Pr\{N(t) = 0\} + f_{\mathbf{Y}}(\mathbf{y}, t \mid N(t) > 0)\Pr\{N(t) > 0\} \\ &= P_0 \prod_{i=1}^{n} \delta\big(y_i\big) + (1 - P_0) f^0_{\mathbf{Y}}(\mathbf{y^0}, t), \end{aligned} \tag{22}$$

where n is the number of state variables.

In the first, discrete, part the Dirac delta spike $\delta(y_i)$ represents the finite probability of the system being at rest, i.e. $y_i = 0$. This probability is, of course, concentrated at the displacement $y_i = 0$. The second, continuous, part i.e. $f^0_{\mathbf{Y}}(\mathbf{y^0}, t)$ is the conditional density, given that at least one impulse occurred.

In the situations when the first, discrete, term in the assumed discrete-continuous density function is predominant, it can be predicted that the probability density function $f_{\mathbf{Y}}(\mathbf{y}, t)$ will be difficult to approximate by the truncated Gram-Charlier expansion. The reason is that the Dirac delta is difficult to approximate in terms of Hermite polynomials. The usual cumulant neglect closure approximations may be then very inaccurate for the unconditional central moments $\kappa_{i_1 \cdots i_n}$.

The Gram-Charlier expansion can be used for the continuous part of the discrete-continuous pdf only, i.e. for the conditional density $f^0_{\mathbf{Y}}(\mathbf{y^0}, t)$. This is equivalent to assuming the closure approximations for the conditional moments.

With the aid of the relationships (identities) between the unconditional moments and conditional moments the final modified cumulant neglect closure approximations are obtained.

The use of modified cumulant neglect closure approximations allows to improve the results for the problems with sparse pulse trains, i.e. with a low mean arrival rate ν [18].

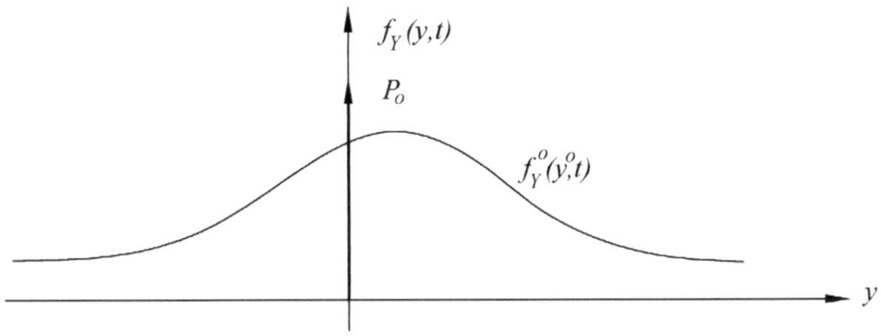

Figure 4. Tentative probability density function of the Poisson-driven response process

3.3. Integro-differential equation governing the response process probability density

The Markov process with continuous-jump sample paths is governed by an **integro-differential Chapman-Kolmogorov equation** (called **master equation** in physics) [9,19]

$$\begin{aligned}\frac{\partial}{\partial t}q_{\{\mathbf{Y}\}}(\mathbf{y},t) &= -\sum_i \frac{\partial}{\partial y_i}[c_i(\mathbf{y},t)q_{\{\mathbf{Y}\}}(\mathbf{y},t)]+\\ &+\int \Big[J_{\{\mathbf{Y}\}}(\mathbf{y}|\mathbf{x},t)q_{\{\mathbf{Y}\}}(\mathbf{x},t) - J_{\{\mathbf{Y}\}}(\mathbf{x}|\mathbf{y},t)q_{\{\mathbf{Y}\}}(\mathbf{y},t)\Big]d\mathbf{x},\end{aligned} \tag{23}$$

$q_{\{\mathbf{Y}\}}(\mathbf{y},t)$ is the joint probability density of the state vector, $c_i(\mathbf{y},t)$ is the drift term of the equation of motion and

$$J_{\{\mathbf{Y}\}}(\mathbf{y}|\mathbf{x},t) = \lim_{\Delta t\to 0}\frac{q_{\{\mathbf{Y}\}}(\mathbf{y},t+\Delta t|\mathbf{x},t)}{\Delta t}. \tag{24}$$

is the **jump probability intensity function**, which is non-zero only if the state vector changes by a finite jump from $\mathbf{x}$ at t to $\mathbf{y}$ at $t+\Delta t$ as $\Delta t \to 0$.

For a system under an external Poisson impulse process excitation, i.e. $\mathbf{b}(\mathbf{Y}(t),t)$ = const. = $\mathbf{b}$, and with a mean arrival rate ν, the insertion of a pertinent jump probability intensity function into Chapman-Kolmogorov equation, followed by integration with respect to $\mathbf{x}$ results in a **Kolmogorov-Feller** integro-differential equation [9].

In the case of a SDOF oscillator, with $\mathbf{b}(\mathbf{Y}(t),t) = const. = \mathbf{b} = [b_1,b_2]^T = [0,b]^T$, the jump probability intensity function is determined as

$$J_{\{\mathbf{Y}\}}(\mathbf{y}|\mathbf{x},t) = \nu\delta(y_1-x_1)\int_{\mathcal{P}}\delta(y_2-(x_2+b_2p))f_P(p)dp \tag{25}$$

The integration of the Chapman-Kolmogorov equation (23) with respect to $\mathbf{x}$ yields a **Kolmogorov-Feller** equation

$$\frac{\partial}{\partial t}q_{\{\mathbf{Y}\}}(\mathbf{y},t) = -\sum_i\frac{\partial}{\partial y_i}[c_i(\mathbf{y},t)q_{\{\mathbf{Y}\}}(\mathbf{y},t)] + \nu\int_{\mathcal{P}}q_{\{\mathbf{Y}\}}(\mathbf{y}-\mathbf{b}p,t)f_P(p)dp - \nu q_{\{\mathbf{Y}\}}(\mathbf{y},t) \tag{26}$$

3.4. Relation to the classical diffusive Markov processes theory

If the impulses mean arrival rate $\nu\to\infty$ and at the same time the impulses magnitudes tend to zero, i.e. the impulse process becomes an infinitely dense train of infinitely small impulses, the velocity response process $\dot{Y}(t)$ reveals jumps at every infinitesimal time interval dt, hence it is discontinuous and consequently the displacement response process $Y(t)$ is non-differentiable. If the impulses have zero-mean values, i.e. $E[P] = 0$, and at the same time their magnitudes tend to zero in such a way that $\nu E[P^2]$ is kept constant, the Poisson impulse process

$$F(t) = \sum_{i=1}^{N(t)} P_i \delta(t - t_i) \tag{27}$$

tends asymptotically to the **Gaussian white noise** and the equations of motion become Itô's stochastic differential equations. In the case of additive excitation

$$d\mathbf{Y}(t) = \mathbf{c}(\mathbf{Y}(t), t)dt + \mathbf{b}dW(t), \quad \mathbf{Y}(0) = \mathbf{y}_0, \tag{28}$$

where $W(t)$ is the **Wiener process**.

In the asymptotic case the generalized Itô's differential rule (13) reduces to the classical Itô's differential rule

$$\begin{aligned} dV(t, \mathbf{Y}(t)) &= \frac{\partial V(t, \mathbf{Y}(t))}{\partial t} dt \\ &+ \sum_{i=1}^{n} \frac{\partial V(t, \mathbf{Y}(t))}{\partial Y_i} \left(c_i \left(\mathbf{Y}(t), t \right) dt + b_i dW(t) \right) + \frac{1}{2} \nu E[P^2] \sum_{i,j=1}^{n} b_i b_j \frac{\partial^2 V(t, \mathbf{Y}(t))}{\partial Y_i Y_j} dt \end{aligned} \tag{29}$$

In the asymptotic case the Kolmogorov-Feller integro-differential equation (26) reduces to the classical Fokker - Planck - Kolmogorov partial differential equation

$$\frac{\partial}{\partial t} q_{\{\mathbf{Y}\}}(\mathbf{y}, t) = -\sum_i \frac{\partial}{\partial y_i} [c_i(\mathbf{y}, t) q_{\{\mathbf{Y}\}}(\mathbf{y}, t)] + \frac{1}{2} \nu E[P^2] \sum_{i,j=1}^{n} b_i b_j \frac{\partial^2 q_{\{\mathbf{Y}\}}(\mathbf{y}, t)}{\partial y_i y_j} \tag{30}$$

4. Non-Markov problems

4.1. Exact conversion of a non-Markov problem into a Markov one

Consider a dynamic system under a renewal process impulse process where the occurrence times of the impulses are driven by a renewal process $R(t)$. The increments $dR(t)$ **are not independent**. The state vector $\mathbf{Y}(t)$ governed by

$$d\mathbf{Y}(t) = \mathbf{c}(\mathbf{Y}(t), t)dt + \mathbf{b}(\mathbf{Y}(t), t))P(t)dR(t), \quad \mathbf{Y}(0) = \mathbf{y}_0, \tag{31}$$

is not a Markov process. Exact conversion of the original non-Markov problem into a Markov one is performed by recasting of the original impulse process (replacement valid with probability 1) as [9,11]

$$\sum_{i,R=1}^{R(t)} P_{i,R} \delta(t - t_{i,R}) = \sum_{i=1}^{N(t)} \rho(N(t_i)) P_i \delta(t - t_i), \tag{32}$$

where $\rho(N(t_i)$ is an **auxiliary, pure jump, zero-one stochastic process**, which selects only some impulses from the train driven by a Poisson process $N(t)$.

As an auxiliary, pure-jump stochastic process consists of negative-exponential distributed "phases", the states of sojourn of the process in these "phases" are **Markov states** $S(t) = 1, 2, ..., m$. The jump process must be defined in such a way: the actual impulse (i.e. the jump in the velocity response $\dot{Y}(t)$) only occurs if there is a transition between some particular Markov states. Other transitions are not associated with a jump in the velocity response.

Original state variables of the dynamic system and the states of the auxiliary jump process are **jointly Markovian**.

Accordingly, the mixed-type joint probability density-discrete distribution function of the continuous-jump processes $Y(t)$, $\dot{Y}(t)$ and of $\mathbf{m}$ Markov states $S(t)$ of the pure jump process is defined as

$$q_j(\mathbf{y},t)d\mathbf{y} = \Pr\{\mathbf{Y}(t) \in (\mathbf{y}+d\mathbf{y}) \wedge S(t)=j\}, \quad j=1,2,...,m \tag{33}$$

This function is governed by the set of integro-differential forward Chapman-Kolmogorov equation

$$\begin{aligned}
&\frac{\partial}{\partial t}q_j(\mathbf{y},t) = \mathcal{K}_j[\mathbf{q}(\mathbf{y},t)]\\
&\mathcal{K}_j[\mathbf{q}(\mathbf{y},t)] = -\sum_{i=1}^{n}\frac{\partial}{\partial y_i}\Big[c_i(\mathbf{y},t)q_j(\mathbf{y},t)\Big]+\\
&\sum_{l=1,l\neq j}^{m}\int_{-\infty}^{\infty}\int_{-\infty}^{\infty}\Bigg[J_{\{\mathbf{Y}\}}(\mathbf{y},j|\mathbf{x},l,t)q_l(\mathbf{x},t) - J_{\{\mathbf{Y}\}}(\mathbf{x},l|\mathbf{y},j,t)q_j(\mathbf{y},t)\Bigg]d\mathbf{x},
\end{aligned} \tag{34}$$

where $\mathcal{K}_j[...]$ is the forward integro-differential Chapman-Kolmogorov operator, $j=1,2,...,m$ and $c_i(\mathbf{y},t)$ are the drift coefficients of the equation of motion.

The explicit integro-differential equations are obtained after the insertion of the suitably determined jump probability intensity function $J_{\{\mathbf{Y}\}}(\mathbf{y},j|\mathbf{x},l,t)$ and subsequent integration with respect to $\mathbf{x}$. The jump probability intensity function defined as

$$J_{\{\mathbf{Y}\}}(\mathbf{y},j|\mathbf{x},l,t) = \lim_{\Delta t\to 0}\frac{\Pr\{\mathbf{Y}(t+\Delta t)=\mathbf{y}, S(t+\Delta t)=j, |\mathbf{X}(t)=\mathbf{x}, S(t)=l\}}{\Delta t} \tag{35}$$

must be determined from the pertinent chain of Markov states.

The non-zero jump probability intensity functions are only defined for $j\neq l$, such that there is a transition in a Markov chain (jump in the auxiliary process) from $S(t)=l$ to $S(t+\Delta t)=j$. These functions assume two different forms: for the transitions in a Markov chain not associated by the jump in the velocity process and those associated by the jump in the velocity process. Let us determine the jump probability intensity function for a single-degree-of-freedom system (two state variables). If there is a transition from $S(t)=l$ to $S(t+\Delta t)=j$, but no actual impulse occurs (no jump in the velocity process)

$$J_{\{\mathbf{Y}\}}(\mathbf{y},j|\mathbf{x},l,t) = \pi(j|l)\delta(y_1-x_1)\delta(y_2-x_2), \tag{36}$$

where

$$\pi(j|l) = \frac{P_{j|l}(\Delta t)}{\Delta t}, \qquad j\neq l, \tag{37}$$

and $P_{j|l}(\Delta t) = \Pr\{S(t+\Delta t)=j\big|S(t)=l\}$ is the transition probability in a pertinent Markov chain. If the transition from $S(t)=l$ to $S(t+\Delta t)=j$, is associated with the actual impulse (the jump in the velocity process $\dot{Y}(t)=Y_2(t)$)

$$J_{\{\mathbf{Z}\}}(\mathbf{y},j|\mathbf{x},l,t) = \pi(j|l)\delta(y_1-x_1)\int_{\mathcal{P}}\delta\Big(y_2-\big(x_2+b(x_1,x_2)p\big)\Big)f_P(p)dp \tag{38}$$

In an Erlang renewal process [20-22] with an integer parameter k the events are every kth Poisson events, hence it must be assumed that the actual impulse, and hence the jump in the velocity variable occurs when the jump is from the state k to 1. All other jumps in the auxiliary process occur from $j-1$ to j ($j=2,...k$), but there are no corresponding impulses. The auxiliary pure jump process and the corresponding Markov chain are shown in the figures 5 and 6.

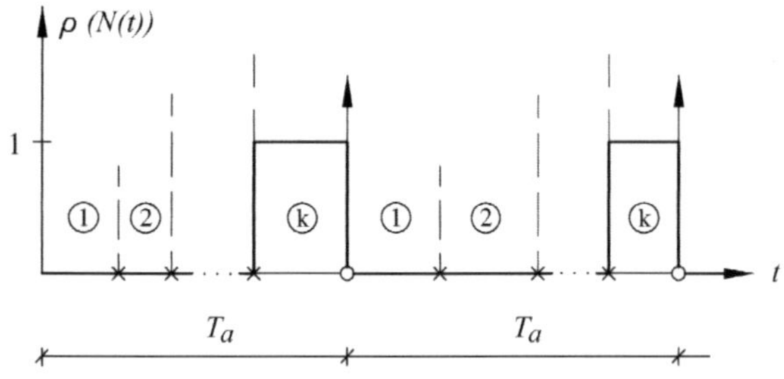

Figure 5. Auxiliary pure jump process

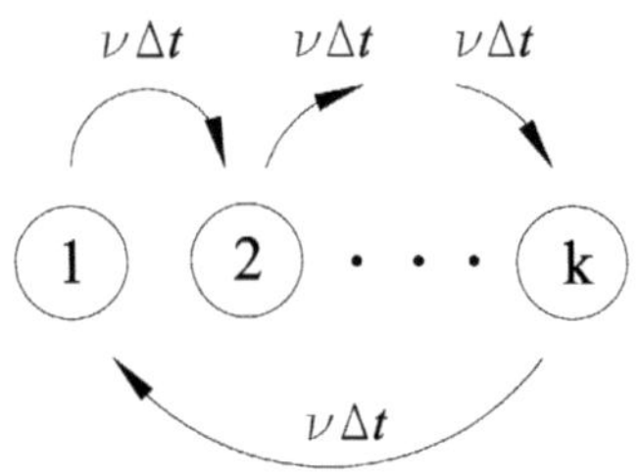

Figure 6. Corresponding Markov chain

The jump probability intensity function is determined as

$$J_{\mathbf{Y}}(y_1, y_2, j|x_1, x_2, l, t) = \begin{cases} \nu\delta(y_1 - x_1) \int\limits_{\mathcal{P}} \delta(y_2 - (x_2 + b(x_1, x_2)p)) f_P(p)dp, \\ \hfill j = 1, l = k. \\ \nu\delta(y_1 - x_1)\delta(y_2 - x_2), \quad j = 2, 3, ..., k, \ l = j - 1 \end{cases} \tag{39}$$

If the excitation is purely external (additive), i.e. $\mathbf{b}(\mathbf{Y}(t), t) = const. = \mathbf{b}$, the integration with respect to $\mathbf{x}$ yields the following equations for the Erlang process with an arbitrary parameter k

$$\begin{aligned} \frac{\partial}{\partial t} q_1(\mathbf{y}, t) &= -\sum_{i=1}^{n} \frac{\partial}{\partial y_i}\Big[c_i(\mathbf{y}, t) q_1(\mathbf{y}, t)\Big] + \nu \int\limits_{\mathcal{P}} q_k(\mathbf{y} - \mathbf{b}p, t) f_P(p)dp - \nu q_1(\mathbf{y}, t) \\ &\vdots \\ \frac{\partial}{\partial t} q_j(\mathbf{y}, t) &= -\sum_{i=1}^{n} \frac{\partial}{\partial y_i}\Big[c_i(\mathbf{y}, t) q_j(\mathbf{y}, t)\Big] + \nu q_{j-1}(\mathbf{y}, t) - \nu q_j(\mathbf{y}, t), \quad j = 2, 3, ...k \end{aligned} \tag{40}$$

The generating equation for moments is derived for the expectations with respect to the joint probability density - discrete distribution function $q_j(\mathbf{y}, t)$

$$E_j\Big[V(\mathbf{Y}(t), t)\Big] = \int\limits_{-\infty}^{\infty} V(\mathbf{y}, t) q_j(\mathbf{y}, t) d\mathbf{y}, \quad j = 1, 2, ..., m. \tag{41}$$

Its time evolution is determined as

$$\begin{aligned} \frac{d}{dt} E_j\Big[V(\mathbf{Y}(t), t)\Big] = \frac{\partial}{\partial t} \int\limits_{-\infty}^{\infty} V(\mathbf{y}(t), t) q_j(\mathbf{y}, t) d\mathbf{y} = \\ E_j\left[\frac{\partial}{\partial t} V(\mathbf{Y}(t), t)\right] + \int\limits_{-\infty}^{\infty} V(\mathbf{y}(t), t) \mathcal{K}_j\Big[\mathbf{q}(\mathbf{y}, t)\Big] d\mathbf{y}. \end{aligned} \tag{42}$$

After the integration by parts, suitable interchange of dummy variables and some rearrangements the generating equation for moments is arrived at in the form [14]

$$\begin{aligned}\frac{d}{dt}E_j\Big[V(\mathbf{Y}(t),t)\Big] &= E_j\left[\frac{\partial}{\partial t}V(\mathbf{Y}(t),t)\right]+\sum_{r=1}^{n}E_j\left[\frac{\partial V(\mathbf{Y}(t),t)}{\partial Y_r}c_r(\mathbf{Y}(t),t)\right]\\ &+\sum_{l=1}^{m}\int_{-\infty}^{\infty}\int_{-\infty}^{\infty}\Big[V(\mathbf{x}(t),t)J_{\{\mathbf{Y}\}}(\mathbf{x},j|\mathbf{y},l,t)q_i(\mathbf{y},t)\\ &\qquad -V(\mathbf{y}(t),t)J_{\{\mathbf{Y}\}}(\mathbf{x},l|\mathbf{y},j,t)q_j(\mathbf{y},t)\Big]d\mathbf{x}d\mathbf{y}, \quad j=1,2,...,m.\end{aligned} \tag{43}$$

As the usual-sense (marginal) joint probability density function is obtained by summation

$$q(\mathbf{y},t)=\sum_{j=1}^{m}q_j(\mathbf{y},t), \tag{44}$$

so is the usual-sense expectation

$$E\Big[V(\mathbf{Y}(t),t)\Big]=\sum_{j=1}^{m}E_j\Big[V(\mathbf{Y}(t),t)\Big] \tag{45}$$

4.2. Moment equations for an Erlang renewal impulse process

For an Erlang renewal impulse process the jump probability intensity function $J_{\{\mathbf{Y}\}}(x_1,x_2,j|y_1,y_2,l,t)$ determined with the aid of the Markov chain (Figure 6) equals

$$J_{\{\mathbf{Y}\}}(x_1,x_2,j|y_1,y_2,l,t)=\begin{cases}\nu\delta(x_1-y_1)\displaystyle\int_{\mathcal{P}}\delta\Big(x_2-(y_2+b(y_1,y_2)p)\Big)f_P(p)dp,\\ \qquad\qquad\qquad\qquad\qquad j=1, l=k.\\ \nu\delta(x_1-y_1)\delta(x_2-y_2),\quad j=2,3,...,k,\ l=j-1\end{cases} \tag{46}$$

After the insertion of the jump probability intensity function into the general generating equation (43) and integration with respect to $\mathbf{x}$ the problem-specific generating equations for moments are obtained as

$$\begin{aligned}\frac{d}{dt}E_1\Big[V(\mathbf{Y}(t),t)\Big] &= E_1\left[\frac{\partial}{\partial t}V(\mathbf{Y}(t),t)\right]+\sum_{r=1}^{n}E_1\left[\frac{\partial V(\mathbf{Y}(t),t)}{\partial Y_r}c_r(\mathbf{Y}(t),t)\right]\\ &+\nu E_k\left[\int_{\mathbf{P}}V\Big(\mathbf{Y}(t)+\mathbf{b}(\mathbf{Y})p,t\Big)f_P(p)dp\right]-\nu E_1\Big[V(\mathbf{Y}(t),t)\Big],\\ \frac{d}{dt}E_j\Big[V(\mathbf{Y}(t),t)\Big] &= E_j\left[\frac{\partial}{\partial t}V(\mathbf{Y}(t),t)\right]+\sum_{r=1}^{n}E_j\left[\frac{\partial V(\mathbf{Y}(t),t)}{\partial Y_r}c_r(\mathbf{Y}(t),t)\right]\\ &+\nu E_{j-1}\Big[V(\mathbf{Y}(t),t)\Big]-\nu E_j\Big[V(\mathbf{Y}(t),t)\Big], \qquad j=2,3,...,k\end{aligned} \tag{47}$$

The explicit equations for first-, second-, third- and fourth-order moments are obtained as

$$\begin{aligned}
\frac{d}{dt}E_1\Big[Z_i(t)\Big] &= E_1\Big[c_i(\mathbf{Z}(t))\Big] + \nu E_k\Big[Z_i(t)\Big] + \nu E[P]E_k\Big[b_i(\mathbf{Z}(t))\Big] - \nu E_1\Big[Z_i(t)\Big] \\
\frac{d}{dt}E_j\Big[Z_i(t)\Big] &= E_j\Big[c_i(\mathbf{Z}(t))\Big] + \nu E_{j-1}\Big[Z_i(t)\Big] - \nu E_j\Big[Z_i(t)\Big], \qquad j = 2,3,...,k
\end{aligned} \tag{48}$$

$$\begin{aligned}
\frac{d}{dt}E_1\Big[Z_i(t)Z_l(t)\Big] &= 2\Big\{E_1\Big[Z_i c_l(\mathbf{Z}(t))\Big]\Big\}_s + \nu E_k\Big[Z_i(t)Z_l(t)\Big] + 2\nu E[P]\Big\{E_k\Big[Z_i b_l(\mathbf{Z}(t))\Big]\Big\}_s \\
&\quad + \nu E[P^2]E_k\Big[b_i(\mathbf{Z}(t))b_l(\mathbf{Z}(t))\Big] - \nu E_1\Big[Z_i(t)Z_l(t)\Big] \\
\frac{d}{dt}E_2\Big[Z_i(t)Z_l(t)\Big] &= 2\Big\{E_j\Big[Z_i c_l(\mathbf{Z}(t))\Big]\Big\}_s + \nu E_{j-1}\Big[Z_i(t)Z_l(t)\Big] - \nu E_j\Big[Z_i(t)Z_l(t)\Big], \quad j = 2,3,...,k
\end{aligned} \tag{49}$$

$$\begin{aligned}
\frac{d}{dt}E_1\Big[Z_i(t)Z_l(t)Z_m(t)\Big] &= 3\Big\{E_1\Big[Z_i Z_l c_m(\mathbf{Z}(t))\Big]\Big\}_s + \nu E_k\Big[Z_i(t)Z_l(t)Z_m(t)\Big] \\
&\quad + 3\nu E[P]\Big\{E_k\Big[Z_i Z_l b_m(\mathbf{Z}(t))\Big]\Big\}_s + 3\nu E[P^2]\Big\{E_k\Big[Z_i b_l(\mathbf{Z}(t)b_m(\mathbf{Z}(t))\Big]\Big\}_s \\
&\quad + \nu E[P^3]E_k\Big[b_i(\mathbf{Z}(t))b_l(\mathbf{Z}(t))b_m(\mathbf{Z}(t)\Big] - \nu E_1\Big[Z_i(t)Z_l(t)Z_m(t)\Big] \\
\frac{d}{dt}E_j\Big[Z_i(t)Z_l(t)Z_m(t)\Big] &= 3\Big\{E_j\Big[Z_i Z_l c_m(\mathbf{Z}(t))\Big]\Big\}_s + \nu E_{j-1}\Big[Z_i(t)Z_l(t)Z_m(t)\Big] \\
&\quad - \nu E_j\Big[Z_i(t)Z_l(t)Z_m(t)\Big] \qquad j = 2,3,...,k
\end{aligned} \tag{50}$$

$$\begin{aligned}
\frac{d}{dt}E_1\Big[Z_i(t)Z_l(t)Z_m(t)Z_n(t)\Big] &= 4\Big\{E_1\Big[Z_i Z_l Z_m c_n(\mathbf{Z}(t))\Big]\Big\}_s + \nu E_k\Big[Z_i(t)Z_l(t)Z_m(t)Z_n(t)\Big] \\
&\quad + 4\nu E[P]\Big\{E_k\Big[Z_i Z_l Z_m b_n(\mathbf{Z}(t))\Big]\Big\}_s \\
&\quad + 6\nu E[P^2]\Big\{E_k\Big[Z_i Z_l b_m(\mathbf{Z}(t)b_n(\mathbf{Z}(t))\Big]\Big\}_s \\
&\quad + 4\nu E[P^3]E_k\Big[Z_i b_l(\mathbf{Z}(t))b_m(\mathbf{Z}(t))b_n(\mathbf{Z}(t)\Big] \\
&\quad + \nu E[P^4]E_k\Big[b_i(\mathbf{Z}(t))b_l(\mathbf{Z}(t))b_m(\mathbf{Z}(t))b_n(\mathbf{Z}(t)\Big] \\
&\quad - \nu E_1\Big[Z_i(t)Z_l(t)Z_m(t)Z_n(t)\Big] \\
\frac{d}{dt}E_j\Big[Z_i(t)Z_l(t)Z_m(t)Z_n(t)\Big] &= 4\Big\{E_j\Big[Z_i Z_l Z_m c_n(\mathbf{Z}(t))\Big]\Big\}_s + \nu E_{j-1}\Big[Z_i(t)Z_l(t)Z_m(t)Z_n(t)\Big] \\
&\quad - \nu E_j\Big[Z_i(t)Z_l(t)Z_m(t)Z_n(t)\Big], \qquad j = 2,3,...,k
\end{aligned} \tag{51}$$

4.3. Example: response of a linear oscillator to an Erlang renewal impulse process with $k = 2$
Consider a linear oscillator under external excitation, governed by the equation

$$\ddot{Y} + 2\zeta\omega\dot{Y} + \omega^2 Y = b\sum_{i=1}^{R(t)} P_i\delta(t - t_i), \tag{52}$$

or

$$\begin{aligned} dZ_1(t) &= Z_2(t)dt \\ dZ_2(t) &= \left(-2\zeta\omega Z_2(t) - \omega^2 Z_1(t)\right) dt + bP(t)dR(t), \end{aligned} \tag{53}$$

hence $c_1(Z_1, Z_2, t) = Z_2$, $c_2(Z_1, Z_2, t) = -\omega^2 Z_1 - 2\zeta\omega Z_2$, $b_1 = 0$, $b_2(Z_1, Z_2, t) = \text{const.} = b$. The equations for the mean values $\mathbf{m}(t) = \left[E_1[Z_1(t)], E_2[Z_1(t)], E_1[Z_2(t)], E_2[Z_2(t)]\right]^T$ are obtained from (47) as

$$\frac{d}{dt}\mathbf{m}(t) = \mathbf{A}_1\mathbf{m}(t) + \mathbf{f}_1(t), \tag{54}$$

where

$$\mathbf{A}_1 = \begin{bmatrix} -\nu & \nu & 1 & 0 \\ \nu & -\nu & 0 & 1 \\ -\omega^2 & 0 & -(2\zeta\omega+\nu) & \nu \\ 0 & -\omega^2 & \nu & -(2\zeta\omega+\nu) \end{bmatrix}, \quad \mathbf{f}_1(t) = \begin{bmatrix} 0 \\ 0 \\ \nu b E[P]\mathcal{P}_2(t) \\ 0 \end{bmatrix}. \tag{55}$$

and $\mathcal{P}_2(t) = \Pr\{S(t) = 2\}$.

According to (45), the usual-sense mean values $E[Z_1]$ and $E[Z_2]$ are

$$E[Z_1] = E_1[Z_1] + E_2[Z_1], \quad E[Z_2] = E_1[Z_2] + E_2[Z_2], \tag{56}$$

hence the equations for these mean values are obtained by pair-wise summation of the equations (54, 55), which yields

$$\begin{aligned} \frac{d}{dt}E[Z_1] &= E[Z_2] \\ \frac{d}{dt}E[Z_2] &= -\omega^2 E[Z_1] - 2\zeta\omega E[Z_2] + \nu b E[P]\mathcal{P}_2(t) \end{aligned} \tag{57}$$

The asymptotic, stationary mean values $\mathbf{m} = \lim_{t\to\infty} \mathbf{m}(t) = \left[E_1[Z_1], E_2[Z_1], E_1[Z_2], E_2[Z_2]\right]^T$ are governed by

$$\mathbf{A}_1\mathbf{m} = -\mathbf{f}_1, \tag{58}$$

with the stationary value of $\mathcal{P}_2(t)$

$$\mathcal{P}_2 = \frac{1}{2} \tag{59}$$

In particular, the stationary solution for the mean value of the displacement response equals (cf. [9])

$$E[Z_1] = \frac{\nu b E[P]}{2\omega^2} \tag{60}$$

The equations for the ordinary second-order moments $\boldsymbol{\mu}(t) = \left[E_1[Z_1^2], E_2[Z_1^2], E_1[Z_1Z_2], E_2[Z_1Z_2], E_1[Z_2^2], E_2[Z_2^2]\right]^T$ are obtained from (47) as

$$\frac{d}{dt}\boldsymbol{\mu}(t) = \mathbf{A}_2\boldsymbol{\mu}(t) + \mathbf{f}_2(t), \tag{61}$$

where

$$\mathbf{A}_2 = \begin{bmatrix} -\nu & \nu & 2 & 0 & 0 & 0 \\ \nu & -\nu & 0 & 2 & 0 & 0 \\ -\omega^2 & 0 & -(2\zeta\omega+\nu) & \nu & 1 & 0 \\ 0 & -\omega^2 & \nu & -(2\zeta\omega+\nu) & 0 & 1 \\ 0 & 0 & -2\omega^2 & 0 & -(4\zeta\omega+\nu) & \nu \\ 0 & 0 & 0 & -2\omega^2 & \nu & -(4\zeta\omega+\nu) \end{bmatrix}, \tag{62}$$

$$\mathbf{f}_2(t) = \begin{bmatrix} 0 \\ 0 \\ \nu b E[P] E_2[Z_1] \\ 0 \\ 2\nu b E[P] E_2[Z_2] + \nu b^2 E[P^2] \mathcal{P}_2(t) \\ 0 \end{bmatrix}. \tag{63}$$

The pair-wise summation of the equations (61 - 63) yields the equations for the usual-sense ordinary second-order moments $E\Big[Z_i(t)Z_l(t)\Big]$

$$\begin{aligned} \frac{d}{dt}E[Z_1^2] &= 2E[Z_1Z_2], \\ \frac{d}{dt}E[Z_1Z_2] &= -\omega^2 E[Z_1^2] - 2\zeta\omega E[Z_1Z_2] + E[Z_2^2] + \nu b E[P] E_2[Z_1], \\ \frac{d}{dt}E[Z_2^2] &= -2\omega^2 E[Z_1Z_2] - 4\zeta\omega E[Z_2^2] + 2\nu b E[P] E_2[Z_2] + \nu b^2 E[P^2] \mathcal{P}_2(t) \end{aligned} \tag{64}$$

In order to obtain the asymptotic, stationary ordinary second-order moments $E\Big[Z_rZ_s\Big]$ the stationary moments $E_2[Z_1]$ and $E_2[Z_2]$ must be determined from the equations (58). The results are

$$E_2[Z_1] = \frac{\nu^2 b E[P](\zeta\omega + \nu)}{\omega^2\left(\omega^2 + 4\nu(\zeta\omega+\nu)\right)}, \qquad E_2[Z_2] = -\frac{\nu^2 b E[P]}{2\left(\omega^2 + 4\nu(\zeta\omega+\nu)\right)}. \tag{65}$$

The stationary solution for the mean-square of the displacement response $E[Z_1^2]$ is

$$E[Z_1^2] = \frac{\nu b^2 E[P^2]}{8\zeta\omega^3} + \frac{\nu^3 b^2 E^2[P](\zeta\omega+\nu)}{\omega^2\left(\omega^2 + 4\nu(\zeta\omega+\nu)\right)}\left(\frac{\zeta\omega+\nu}{\omega^2} - \frac{1}{4\zeta\omega}\right) \tag{66}$$

Assume the data: $\omega = 1\ [s^{-1}]$, $\zeta = 0.05$, $b = 1\ [kg^{-1}]$, $\nu = 20\ [s^{-1}]$, and assume that the impulses magnitudes P are Rayleigh distributed with parameter $\sigma = 0.1\ [\frac{kg\cdot m}{s}]$. Hence $E[P] = \sigma\sqrt{\frac{\pi}{2}} = 0.12533\ [\frac{kg\cdot m}{s}]$, $E[P^2] = 2\sigma^2 = 0.02\ [\frac{kg^2\cdot m^2}{s^2}]$.

Variance of the stationary displacement response

$$var(Z_1) = E[Z_1^2] - E^2[Z_1] = 2.178 - 1.2533^2 = 0.607[\mathrm{m}^2] \tag{67}$$

Mean value and variance of the response to Erlang renewal impulse process excitation were determined in [9] with the aid of the method of state vector augmentation. The above value of $var(Z_1)$ is in perfect agreement with the numerical results given in [9].

5. Conclusions

The review of the methods for determination of the response of mechanical dynamic systems to Poisson and non-Poisson impulse process stochastic excitations is presented. Stochastic differential and integro-differential counterparts of the usual differential equations of motion are introduced. For systems driven by Poisson impulse process the following tools of the theory of non-diffusive Markov processes are presented: the generalized Itô's differential rule and the forward integro-differential Chapman-Kolmogorov equation. It is shown how these tools allow to derive the differential equations for response moments and the integro-differential equation (Kolmogorov-Feller equation) governing the probability density of the response. The relation of Poisson impulse process problems to the theory of diffusive Markov processes is also discussed. For systems driven by a class of non-Poisson (Erlang renewal) impulse processes an exact conversion of the original non-Markov problem into a Markov one is based on the appended Markov chain corresponding to the introduced auxiliary pure jump stochastic process. The derivation of the set of integro-differential equations for response probability density and also a moment equations technique are based on the forward integro-differential Chapman-Kolmogorov equation. An illustrating numerical example is also included.

References

[1] Lindgren G 1981 Jumps and bumps on random roads. *J. Sound and Vibration* **78** pp 383-395
[2] Roberts J B 1966 On the response of a simple oscillator to random impulses *J. Sound and Vibration* **4** pp 51-61
[3] Tung C C 1967 Random response of highway bridges to vehicle loads *J. Engrg. Mech. Div. ASCE* **93** pp 9-94
[4] Tung C C 1969 Response of highway bridges to renewal traffic loads *J. Engrg. Mech. Div. ASCE* **95** pp 41-57
[5] Cornell C A 1964 Stochastic process models in structural engineering *Technical Report Dept. of Civil Engrg* Stanford vol 34
[6] Merchant D H 1964 Stochastic model of wind gusts *Technical Report Dept. of Civil Engrg* Stanford vol 48
[7] Racicot R and Moses F 1972 Filtered Poisson process for random vibration problems *J. Engrg. Mech. Div. ASCE* **98** pp 159-176
[8] Nielsen S R K Iwankiewicz R and Skjaerbaek P S 1995 Moment equations for non-linear systems under renewal-driven random impulses with gamma-distributed interarrival times *Proc. IUTAM Symposium on Advances in Nonlinear Mechanics* ed Naess A and Krenk S Trondheim Norway Kluwer pp 331-340
[9] Iwankiewicz R and Nielsen S R K 1999 *Advanced methods in stochastic dynamics of non-linear systems* Aalborg University Press Denmark
[10] Iwankiewicz R 2002 Dynamic response of non-linear systems to random trains of non-overlapping pulses *Meccanica* **37** pp 167-178
[11] Iwankiewicz R 2005 Equation for probability density of the response of a dynamic system to Erlang renewal random impulse processes *Proc. 12th IFIP WG 7.5 Working Conference on Reliability and Optimization of Structural Systems* ed. Soerensen J D and Frangopol D M Aalborg Denmark Taylor and Francis pp 107-113
[12] Tellier M and Iwankiewicz R 2005 Response of linear dynamic systems to non-Erlang renewal impulses: stochastic equations approach *Probabilistic Engineering Mechanics* **20** pp 281-295
[13] Iwankiewicz R 2008 Equations for probability density of response of dynamic systems to a class of non-Poisson random impulse process excitations *Probabilistic Engineering Mechanics* **23** pp 198-207
[14] Iwankiewicz R 2014 Response of Dynamic Systems to Renewal Impulse Processes: Generating Equation for Moments based on the Integro-Differential Chapman-Kolmogorov Equations *Probabilistic Engineering Mechanics* **35** pp 52-56
[15] Tylikowski A 1982 Vibration of a harmonic oscillator due to a sequence of random impacts *Proc. of the Inst. of Foundations of Machines Construction* Technical University of Warsaw Poland
[16] Snyder D L 1975 *Random point processes* J. Wiley New York
[17] Snyder D L Miller M I 1991 *Random point processes in time and space* Springer-Verlag New York
[18] Iwankiewicz R Nielsen S R K and Thoft-Christensen P 1990 Dynamic response of non-linear systems to Poisson-distributed pulse trains: Markov approach *Structural Safety* **8** pp 223-238
[19] Gardiner C W 1985 *Handbook of stochastic methods for physics, chemistry and the natural sciences* Springer-Verlag
[20] Saaty T 1961 *Elements of queueing theory* McGraw-Hill New York
[21] Cox D R 1962 *Renewal theory* Methuen London
[22] Gross D and Harris C M 1985 *Fundamentals of queueing theory* J. Wiley New York

Quasi-stationary mechanics of elastic continua with bending stiffness wrapping on a pulley system

S Kaczmarczyk, S Mirhadizadeh

School of Science and Technology, University of Northampton, St. George's Avenue, Northampton NN2 6JD, UK

stefan.kaczmarczyk@northampton.ac.uk

Abstract. In many engineering applications elastic continua such as ropes and belts often are subject to bending when they pass over pulleys / sheaves. In this paper the quasi-stationary mechanics of a cable-pulley system is studied. The cable is modelled as a moving Euler-Bernoulli beam. The distribution of tension is non-uniform along its span and due to the bending stiffness the contact points at the pulley-beam boundaries are not unknown. The system is described by a set of nonlinear ordinary differential equations with undetermined boundary conditions. The resulting nonlinear Boundary Value Problem (BVP) with unknown boundaries is solved by converting the problem into the 'standard' form defined over a fixed interval. Numerical results obtained for a range of typical configurations with relevant boundary conditions applied demonstrate that due to the effects of bending stiffness the angels of wrap are reduced and the span tensions are increased.

1. Introduction

Elastic continua such as ropes and belts are used in lifting and hoisting installations as suspension/ power transmission members and mass compensation means. In these applications elastic continua wrap around rotating sheaves and pulleys. Their behaviour has been studied extensively. One approach is to model the elastic continua without taking into consideration the contact mechanics at the sheave/ pulley boundary. The continua are then modelled without taking into consideration of thir bending stiffness as a string [1-6]. Alternatively, a straight beam model is used to accommodate the bending stiffness effects [7-10]. It is then assumed in the analysis that they approach the sheave / pulley along a straight tangent line. However, due to their finite bending stiffness a transient region exists in the vicinity of the contact point on the sheave/ pulley. At the contact/ boundary point their curvature matches the curvature of the sheave/ pulley, and then as the sections move away from this point, the curvature changes continuously [11-13]. Bending of moving continua around the sheaves/ pulleys reduces the wrap angles, tend to change the natural frequencies and results in a non-trivial steady-state configuration about which the system can vibrate [14].

In this paper the cable span is modelled as an Euler-Bernoulli beam. The distribution of tension is non-uniform along its span and due to the bending stiffness the contact points at the pulley-beam boundaries are not unknown. The system is described by a set of nonlinear ordinary differential equations with undetermined boundary conditions. The resulting nonlinear Boundary Value Problem

(BVP) with unknown moving boundaries is solved by converting the problem into the 'standard' form defined over a fixed interval.

2. Mathematical model

A quasi-stationary motion of a system comprising an axially moving elastic one-dimensional continuum with bending stiffness divided into $p=1,2,3,\ldots,P$ sections of slowly-varying length constrained by discrete elements such as rigid-body masses and rotating pulleys can be described by the following nonlinear Boundary Value Problem (BVP) with unknown boundaries

$$\mathbf{y}'_p = \mathbf{f}_p\left(s_p, \mathbf{y}_p; \tau\right), \quad 0 \le s_p \le \hat{L}_p(\tau) \tag{1}$$

$$\mathbf{g}_p\left[\mathbf{y}_p(0), \mathbf{y}_p\left(\hat{L}_p\right)\right] = 0 \tag{2}$$

where $\mathbf{y}_p$, $\mathbf{f}_p$, $\mathbf{g}_p$ have n components, s_p represents an Eulerian coordinate and $\tau = \varepsilon t$ is a *slowing time* [15]. The formulation of the problem is described in what follows.

2.1. Governing equations

The continuum is modelled as an Euler-Bernoulli beam in vertical configuration moving at speed v. A free-body diagram of a segment of length ds of the beam section p is shown in figure 1, where s (after dropping the subscript p) represents the arc length and is used as an Eulerian coordinate measured along the deformed beam. The Euler-Bernoulli beam model requires that the bending moment M and the shear force Q are given as

$$M = EI\kappa \tag{3}$$

$$Q = \frac{dM}{ds} = EI\frac{d\kappa}{ds} \tag{4}$$

respectively, where EI is the bending stiffness and κ represents the curvature. The curvature is equal to the reciprocal of the radius of curvature ρ can be expressed as the rate of change of the slope angle φ as

$$\kappa = \frac{1}{\rho} = \frac{d\varphi}{ds} \tag{5}$$

The equilibrium of forces in the normal direction and the tangential direction, respectively, yield the following equations to determine the unknown curvature $\kappa(s;\tau)$, tension $T(s;\tau)$ and the slope angle $\varphi(s;\tau)$

$$\kappa'' - \frac{1}{EI}\left(T - mv^2\right)\kappa + \frac{mg}{EI}\sin\varphi = 0 \tag{6}$$

$$T' - mvv' + EI\kappa\kappa' + mg\cos\varphi = 0 \tag{7}$$

where g is the acceleration of gravity, m is the mass per unit length of the beam, $(\ )'$ denotes differentiation with respect to s and the unknown quantities are defined over the interval $0 \le s \le \hat{L}(\tau)$, where the slowly varying boundary $\hat{L}(\tau)$ represents the total arc length of the beam section and is also unknown. Furthermore, the Cartesian coordinates of cable particles are expressed as

$$\frac{\partial x}{\partial s} = \cos\varphi, \quad \frac{\partial y}{\partial s} = \sin\varphi \tag{8}$$

Equations (5-8) can be brought into the form (1) by denoting

$$\mathbf{y} = \left[\kappa, \kappa', T, \hat{L}, \varphi, x, y \right]^T \tag{9}$$

with the function $\mathbf{f}$ defined as the following vector

$$\mathbf{f} = \left[\kappa', \quad \frac{1}{EI}\left(T - mv^2\right)\kappa - \frac{mg}{EI}\sin\varphi, \quad mvv' - EI\kappa\kappa' - mg\cos\varphi, \quad 0, \quad \kappa, \quad \cos\varphi, \quad \sin\varphi \right]^T \tag{10}$$

The boundary conditions (2) are then be formulated according to the span configuration accordingly.

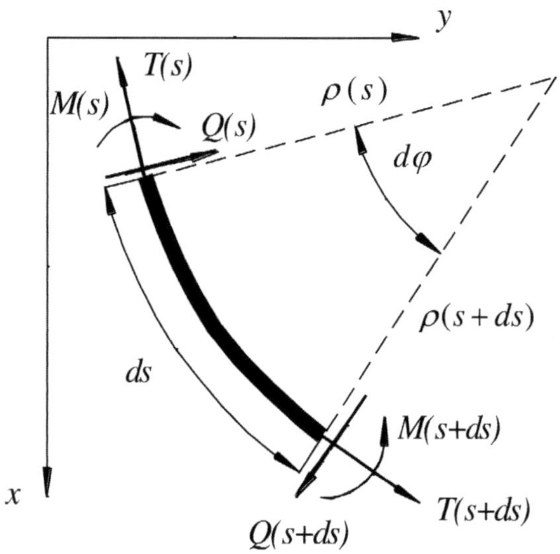

Figure 1. Segment of the beam section.

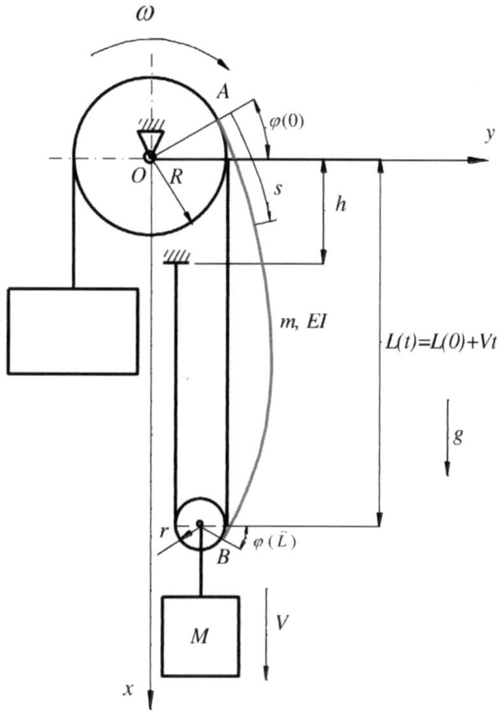

Figure 2. Cable - pulley system.

2.2. The solution strategy

In order to solve equations (1-2) the BVP with the unknown boundary is reformulated into 'standard' form [16] defined over a fixed interval. This is accomplished by introducing the following nondimensional variables

$$\tilde{s} = \frac{s}{\hat{L}}, \quad \tilde{x} = \frac{x}{\hat{L}}, \quad \tilde{y} = \frac{y}{\hat{L}}, \quad \tilde{\kappa} = \hat{L}\kappa, \quad \tilde{T} = \frac{T\hat{L}^2}{EI}, \tilde{v} = \frac{v}{c_0} \tag{11}$$

where $c_0 = \sqrt{T_0/m}$ with T_0 representing the tension, assumed to be known, at the start point or at the end point of the solution interval. Substituting (11) into equations (6-7) yields

$$\tilde{\kappa}'' - \left(\tilde{T} - \beta^2 \hat{L}^2 \tilde{v}^2\right)\tilde{\kappa} + \frac{mg}{EI}\hat{L}^3 \sin\varphi = 0 \tag{12}$$

$$\tilde{T}' - \beta^2 \hat{L}^2 \tilde{v}\tilde{v}' + \tilde{\kappa}\tilde{\kappa}' + \frac{mg}{EI}\hat{L}^3 \cos\varphi = 0 \tag{13}$$

where $\beta = \sqrt{T_0/EI}$ and $(\ )'$ denotes now differentiation with respect to the nondimensional variable $\tilde{s}$ defined over the fixed interval $0 \le \tilde{s} \le 1$. Equations (9-10) assume then the following form

$$\mathbf{y} = \left[\tilde{\kappa}, \tilde{\kappa}', \tilde{T}, \hat{L}, \varphi, \tilde{x}, \tilde{y}\right]^T \tag{14}$$

$$\mathbf{f} = \left[\tilde{\kappa}',\ \ \left(\tilde{T} - \beta^2 \hat{L}^2 \tilde{v}^2\right)\tilde{\kappa} - \frac{mg}{EI}\hat{L}^3 \sin\varphi,\ \ \beta^2 \hat{L}^2 \tilde{v}\tilde{v}' + \tilde{\kappa}\tilde{\kappa}' - \frac{mg}{EI}\hat{L}^3 \cos\varphi,\ \ 0,\ \ \tilde{\kappa},\ \ \cos\varphi,\ \ \sin\varphi\right]^T \tag{15}$$

respectively. In this formulation the problem can be solved by using a standard BVP solver.

3. Numerical example and results

The solution is demonstrated for span *AB* of the cable – pully system shown in figure 2. It is assumed that no slip occurs at the cable – pulley interface. The following boundary conditions expressed in non-dimensional coordinates are applied

$$\kappa(0) = -\frac{\hat{L}(0)}{R},\ \kappa(\hat{L}) = -\frac{\hat{L}(1)}{r},\ \tilde{T}(1) = \frac{\hat{L}^2(1)}{EI}T_0,\ x(0) = -\frac{R}{\hat{L}(0)}\sin\varphi(0),\ y(0) = \frac{R}{\hat{L}(0)}\cos\varphi(0)$$
$$x(\hat{L}) = \frac{1}{\hat{L}(1)}\left[L - r\sin\varphi(1)\right],\ y(\hat{L}) = \frac{1}{\hat{L}(1)}\left[R + r\left(\cos\varphi(1) - 1\right)\right] \tag{16}$$

where $T_0 = \dfrac{Mg}{2\cos\varphi(\hat{L})}$ is the tension at $s = \hat{L}$.

The BVP described by equations (1-2) and equations (14-15) with the boundary conditions (16) is solved over a time interval of 5 s using the parameters shown in Table 1. Sample results are presented in figures 3–5. Figure 3 shows the slope angle at the contact points at $s = 0$ and at $s = \hat{L}$, respectively. It is evident that the bending stiffness decreases the angle of wrap. At the boundary with larger curvature the influence of bending stiffness is more prominent. The curvature changes over the span length as demonstrated in figure 4. The distribution curves of tension along the span are shown in figure 5. It is evident that the tension is increased with bending stiffness.

Table 1. Fundamental parameters of the system

Parameter	Value	Unit
M	100	kg
m	0.343; 2.15	kg/m
EI	1.6; 15	Nm^2
V	1.5	m/s
$L(0)$	2	m
R	0.75	m
r	0.25	m

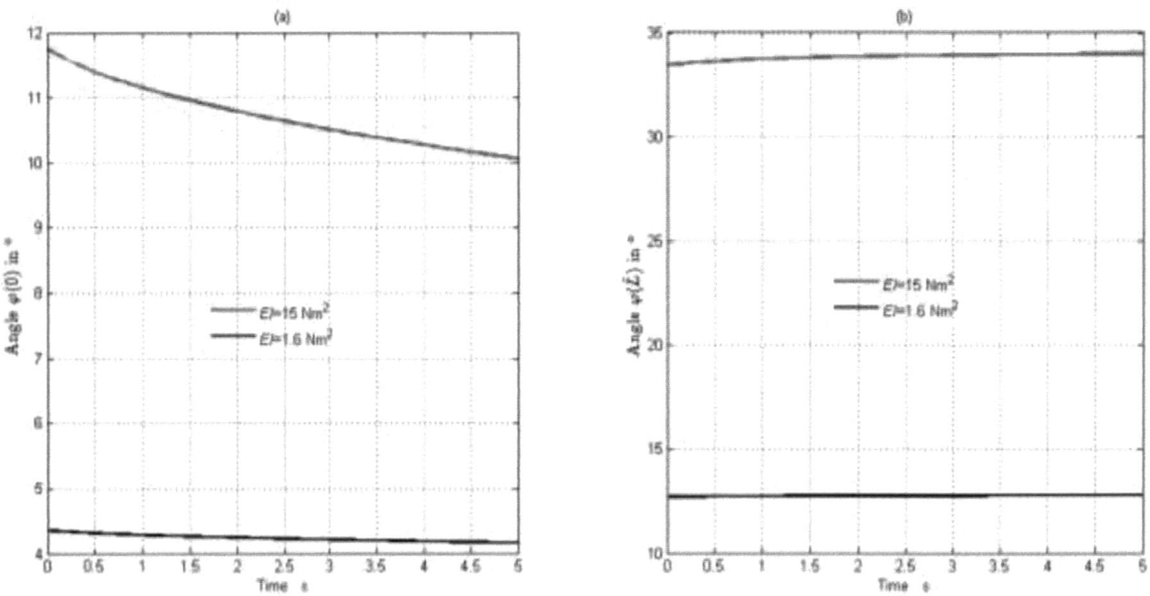

Figure 3. Slope angles at contact points (a) $s = 0$, (b) $s = \hat{L}$.

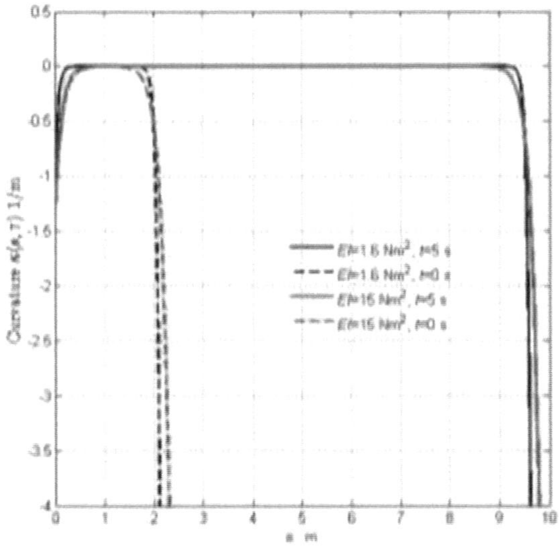

Figure 4. Curvature $\kappa(s;\tau)$.

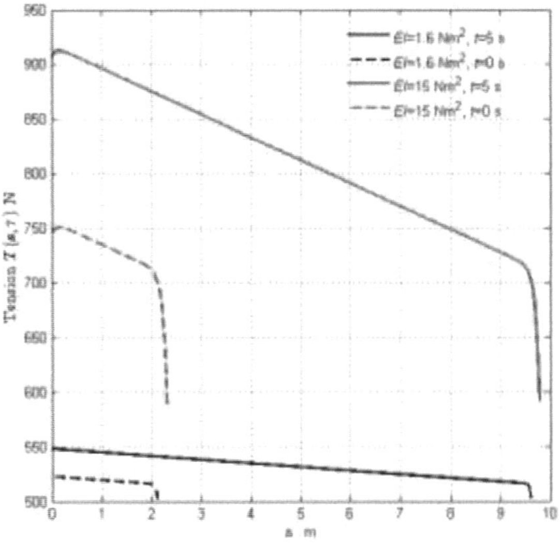

Figure 5. Tensions $T(s;\tau)$.

4. Conclusions

Axially moving elastic continua such as belts, cables and ropes are subject to bending when being wrapped on rigid pulleys/ sheaves. The continua can be represented as a moving Euler-Bernoulli beam. The quasi-stationary mechanics of the system is then described by a nonlinear BVP with unknown slowly varying boundaries at the contact points. The model accommodates the weight of the cable and centrifugal effects. The BVP can be solved by converting the problem into the 'standard' form defined over a fixed interval. The solution yields the span shape, curvature values, slope angles and the distribution of tension along the span. Due to the bending stiffness the wrap angles are decreased and the distribution of tension is nonuniform. The bending stiffness results in the tension being increased which affects useful life of the system.

References

[1] Sandilo S H and van Horssen W T On Variable Length Induced Vibrations of a Vertical String 2013 *Journal of Sound and Vibration* **333**(11) pp 2432-2449

[2] Kaczmarczyk S and Picton P D The prediction of nonlinear responses and active stiffness control of moving slender continua subjected to dynamic loadings in vertical host structures 2013 *The International Journal of Acoustics and Vibration* **18**(1) pp 39-44

[3] Kaczmarczyk S The nonstationary, nonlinear dynamic interactions in slender continua deployed in high-rise vertical transportation systems in the modern built environment 2012 *Journal of Physics: Conference Series* **382** 012037

[4] Zhu W D, Song X K and Zheng N A Dynamic stability of a translating string with a sinusoidally varying velocity 2011 *Transactions of the ASME Journal of Applied Mechanics* **78** 061021

[5] Chen L-Q, Zhang W and Zu J W Nonlinear dynamics for transverse motion of axially moving strings 2009 *Chaos, Solitons & Fractals* **40** pp 78-90

[6] Zhu W D and Zheng N A Exact response of a translating string with arbitrarily varying length under general excitation 2008 *Transactions of the ASME Journal of Applied Mechanics* **75** 031003

[7] Kaczmarczyk S and Mirhadizadeh S Modelling simulation and experimental validation of nonlinear dynamic interactions in an aramid rope system 2014 *Applied Mechanics and Materials, Special Issue: Dynamics and Control of Technical Systems* **706** pp 117-127

[8] Sandilo S H and van Horssen W T On boundary damping for an axially moving tensioned beam 2012 *Journal of Vibration and Acoustics* **134**(1) pp 110051-58

[9] Zhu W D, Ren H and Xiao C A nonlinear model of a slack cable with bending stiffness and moving ends with application to elevator traveling and compensation cables 2011 *Transactions of the ASME Journal of Applied Mechanics* **78** 041017.

[10] Zhu W D and Xu G Y Vibration of elevator cables with small bending stiffness 2003 *Journal of Vibration and Acoustics* **263** pp 679-699

[11] Глушко М Ф Стальные подъемные канаты 1966 (Киев: Техника)

[12] Benson R C Stiff elastic tape wrapped onto a drum 1998 *Transactions of the ASME Journal of Applied Mechanics* **65** pp 870-874

[13] Kong L and Parker R G Steady mechanics of belt-pulley systems 2005 *Transactions of the ASME Journal of Applied Mechanics* **72** pp 25-34

[14] Kong L and Parker R G Vibration of an axially moving beam wrapping on fixed pulleys 2005 *Journal of Sound and Vibration* 280 pp 1066-1074

[15] Kaczmarczyk S The passage through resonance in a catenary-vertical cable hoisting system with slowly varying length 1997 *Journal of Sound and Vibration* **208**(2) pp 243 – 269

[16] Ascher U and Russell R D Reformulation of boundary value problems into "standard" form 1981 *SIAM Review* **23**(2) pp 239-254

Torsional vibrations of helically buckled drill-strings: experiments and FE modelling

M Kapitaniak, V V Hamaneh, M Wiercigroch

E-mail: m.kapitaniak@abdn.ac.uk, vahid.vaziri@abdn.ac.uk, m.wiercigroch@abdn.ac.uk
Centre for Applied Dynamics Research, School of Engineering, University of Aberdeen, Aberdeen AB24 3UE, UK

Abstract. This paper presents investigations of a complex drill-string vibrations on a novel experimental rig, developed by the Centre for Applied Dynamics Research at the University of Aberdeen. The rig is capable of exhibiting of all major types of drill-string vibrations, including torsional, axial and lateral modes. The importance of this work lies in the fact, that the experimental rig utilizes real industrial drill-bits and rock samples, which after careful identification of Torque On Bit (TOB) speed curves, allows to use an equivalent friction model to accommodate for both frictional and cutting components of the bit-rock interactions. Moreover, the proposed Finite Element model, after a careful calibration, is capable of replicating experimental results, for the prebuckled configuration of the drill-string. This allows us to observe the effect of winding and unwinding of the helical deformation during stick-slip motion.

1. Introduction

The problem of drill-string dynamics has gathered a significant interest in the recent times, due to the importance of identification, understanding and solving problems associated with drilling for oil and gas exploration and extraction. Foremost of those are vibrations of different types, such as torsional, lateral and axial ones, that lead to premature wear and damage of the equipment, that result in fatigue and catastrophic failures. In this paper, we focus our attention on stick-slip phenomenon, which is a self excited vibration mode, happening due to nonlinear characteristics of drill-bit rock interactions, that is of particular interest to the drilling community, due to its negative effect on drilling efficiency [2]. Stick-slip has been analysed in the past using mainly low-dimensional model, based on the torsional pendulum [5, 9, 11, 12, 10] or Finite Element (FE) based models [1, 4, 8].

The structure of this paper is as follows. In Section 2 we describe briefly the experimental setup, which is followed by the description of the FE model. Following, in Section 3 we demonstrate experimental verification of the proposed model and the paper finishes with the conclusions given in Section 4.

2. Experimental rig for investigations of drill-string vibrations

The main objective of the experimental rig developed in the Centre for Applied Dynamics Research, University of Aberdeen, is to provide a comprehensive test bed for investigation of different types of undesired vibrations associated with the industrial drilling for oil and gas exploration and extraction. These include torsional, lateral and axial vibrations. Besides, the

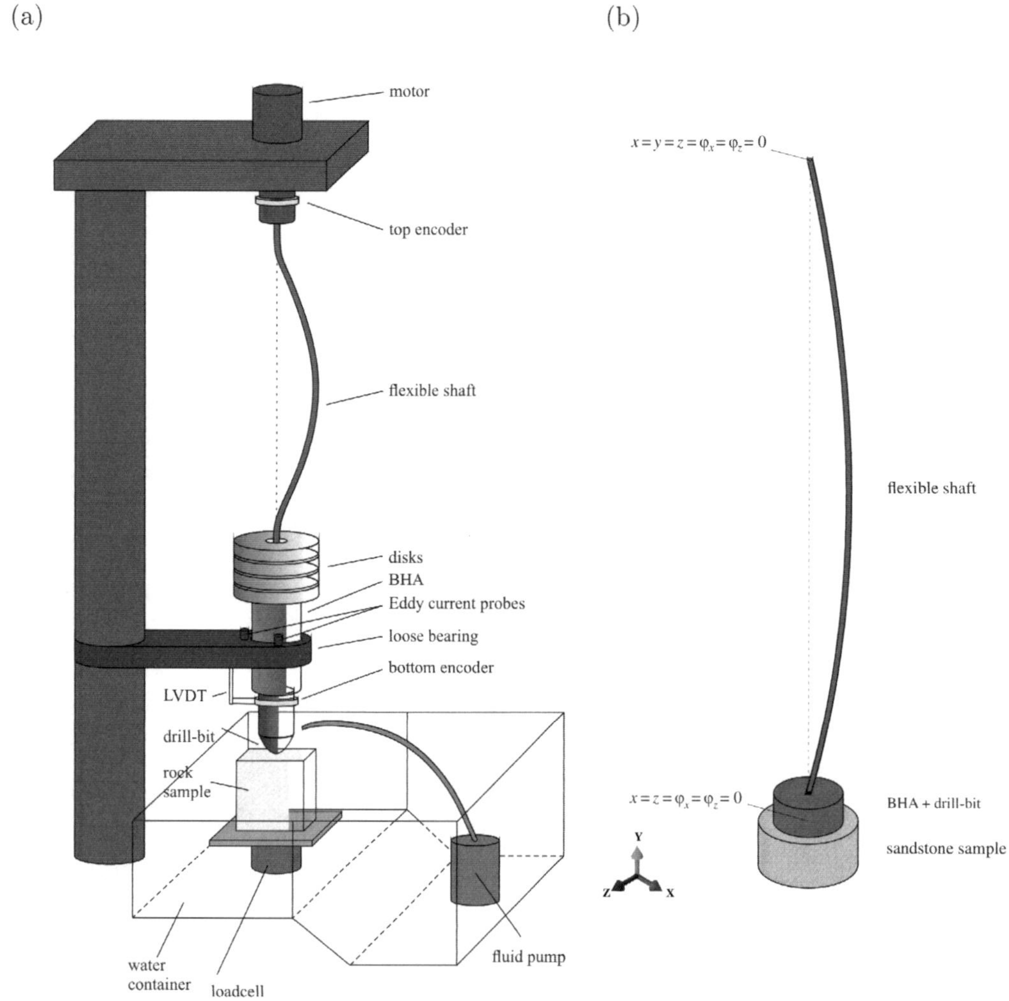

Figure 1. (a) Schematic diagram of the experimental rig, where the main components of the system are: sensors (top and bottom encoders, LVDT, eddy current probes and 4-component load-cell), electric motor, flexible shafts, WOB disks, BHA, drill-bit and rock sample. (b) Configuration of the FE model for pre-buckled configuration, showing all the boundary conditions applied at top and bottom nodes.

rig is capable of providing an insight into the interactions between the real industrial drill-bits, used in the setup and various rock samples, which include for example: sandstone, granite and limestone. The schematic of the rig is shown in Fig. 1 (a), where the main components of the setup can be grouped as follows:

- drill-string composed of flexible/rigid shaft, Bottom Hole Assembly (BHA), Weight On Bit (WOB) disks, and the drill-bit,
- rock samples and cutting fluid circulation system and
- sensors, instrumentation and Data Acquisition System (DAQ).

The purpose of the current study is to develop a numerical model, which is carefully calibrated based on the experimental setup, described in detail in the following sections. The details about the rig description and the instrumentation used can be found in [7, 6].

As described in [7] , the flexible shaft exhibits different properties depending on direction (e.g. low lateral and high torsional stiffness), and thereby it is required to describe it as an anisotropic material. An orthotropic material has in general different properties along 3 principal axes [3]. Therefore, 9 independent materials constants are required to form elastic matrix **D** required to describe the stress-strain relationship, $\sigma = \boldsymbol{D}\varepsilon$. These are three Young's moduli E_1, E_2, E_3, associated with principal axes 1, 2, 3, respectively, three shear moduli G_{12}, G_{13}, G_{23} and three Poisson's ratios ν_{12}, ν_{13}, ν_{23} for planes 1-2, 1-3, 2-3, respectively. The relation between the strain and the stress, $\varepsilon = \boldsymbol{D}^{-1}\sigma$, can be expressed in matrix form as:

$$\begin{pmatrix} \varepsilon_{11} \\ \varepsilon_{22} \\ \varepsilon_{33} \\ \gamma_{12} \\ \gamma_{13} \\ \gamma_{23} \end{pmatrix} = \begin{pmatrix} \frac{1}{E_1} & \frac{-\nu_{21}}{E_2} & \frac{-\nu_{31}}{E_3} & 0 & 0 & 0 \\ \frac{-\nu_{12}}{E_1} & \frac{1}{E_2} & \frac{-\nu_{32}}{E_3} & 0 & 0 & 0 \\ \frac{-\nu_{13}}{E_1} & \frac{-\nu_{23}}{E_2} & \frac{1}{E_3} & 0 & 0 & 0 \\ 0 & 0 & 0 & \frac{1}{G_{12}} & 0 & 0 \\ 0 & 0 & 0 & 0 & \frac{1}{G_{13}} & 0 \\ 0 & 0 & 0 & 0 & 0 & \frac{1}{G_{23}} \end{pmatrix} \begin{pmatrix} \sigma_{11} \\ \sigma_{22} \\ \sigma_{33} \\ \sigma_{12} \\ \sigma_{13} \\ \sigma_{23} \end{pmatrix}. \tag{1}$$

For details of calibration of the torsional properties of the FE model of the flexible shaft, please refer to [7]. In order to evaluate the bending characteristics of the flexible shaft, which are of high importance for modelling of the prebuckled configuration described in the previous section, we perform a series of simple experiments. Therefore, the length of the flexible shaft is limited to 20 cm, by clamping the flexible shaft at the appropriate length, as shown in Fig. 2(a).

As a result, we have a system that is equivalent to cantilever beam, with fixed-free boundary conditions. As can be seen in Fig. 2(a), the flexible shaft on the right hand side has an end fitting (of mass 0.091 kg), which coupled with the weight of the flexible shaft section, results in an end deflection. This state of the flexible shaft is considered as an initial reference configuration. Furthermore, to test bending mechanical properties, we attach to the end of the flexible shaft a number of disks of equal mass (100 g), and observe the resultant end deflection. This is shown in Fig. 2(b)-(c), which correspond to loading with 1-8 disks, respectively. Moreover, the cable behaves in a consistent way, regardless the direction of bending, as can be seen in Fig. 2(d), which presents experimental results of end deflection for both transversal directions, as a function of number of disks attached at flexible shafts end. Clearly, the symmetry in transversal direction is observed. The measurements of experimental end deflection, have been conducted using image recognition technique, which shows that both transversal deflections are consistent, as shown in Fig. 2(d).

One of the critical steps in the modelling of drill-string dynamics is the mathematical description of the bit-rock interactions. Quantitatively, these interactions are represented in terms of Torque On Bit (TOB) versus bit velocity curves, which need to be appropriately modelled. For this purpose, we will follow the approach described [7], where the main idea behind it, is to use suitable effective friction coefficient (slip-rate dependent) that captures the main phenomena observed in our drilling tests, namely, cutting and friction between the drill-bit and the rock sample. According to drilling tests conducted in [7] a presence of the Stribeck effect [13] can be observed, therefore the friction model to be used for the mathematical description of the bit-rock interaction is

$$\mu(r, \omega_b) = \mu_k + (\mu_s - \mu_k)\, e^{-d_c r \omega_b} + \mu_{str} r \omega_b, \tag{2}$$

where μ_k, μ_s, μ_{str} and d_c represent the kinetic, static, Stribeck friction coefficient and the decay rate respectively. Based on that, as described in [7], the resultant TOB function is given by

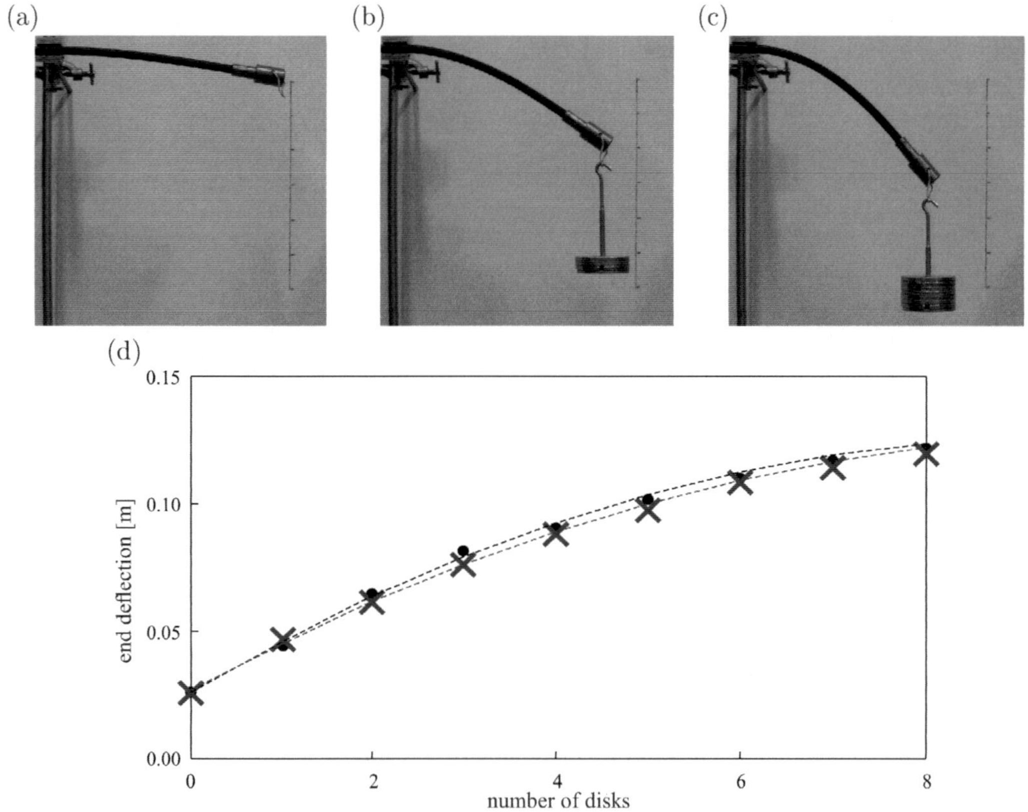

Figure 2. Experiments to determine mechanical characteristics of the flexible shaft. Examples of static bending of a flexible shaft (a)-(c) subjected to different loading conditions and (d) corresponding end deflections of a flexible shaft, subjected to different loading (number of disks) for both transversal directions (denoted with dots and crosses).

$$T_{b,sl}(\omega_b) = \begin{cases} \dfrac{2}{3}\lambda_s W_b, & \omega_b = 0, \\ \dfrac{2}{3}\lambda_k W_b + \dfrac{2W_b(\lambda_s - \lambda_k)}{\lambda_d^3 \omega_b^3}\left(2 - e^{-\lambda_d \omega_b}\left(\lambda_d^2 \omega_b^2 + 2\lambda_d \omega_b + 2\right)\right), & \omega_b > 0, \\ +\dfrac{1}{2} W_b \lambda_{str} \omega_b & \end{cases} \tag{3}$$

where $\lambda_s = \mu_s R$, $\lambda_k = \mu_k R$ and $\lambda_d = d_c R$, $\lambda_{str} = \mu_{str} R^2$, ω_b and R represent angular velocity and radius of the drill-bit. Based on the experimental results, a curve fitting of the experimental TOB (See Fig. 3 (a)) to the Eq. (3) was performed, which allows to obtain necessary parameters for equivalent friction coefficients (see Fig. 3 (b)) corresponding to torque curves for each WOB case. All the necessary parameters to construct experimental torque curves are given in Table 1.

Table 1. Parameters of the equivalent friction coefficients describing bit-rock interactions.

W_b [kN]	λ_s	λ_k	λ_d	λ_{str}
2.19	0.0059	0.0051	0.3091	1.4890E-05
2.06	0.0068	0.0051	0.3678	5.9844E-06
1.88	0.0063	0.0044	0.1219	1.7256E-05
1.57	0.0058	0.0044	0.1770	2.3248E-05
1.43	0.0063	0.0043	0.1085	2.0700E-05
1.22	0.0059	0.0046	0.1298	1.8227E-05
1.10	0.0058	0.0036	0.0695	3.4923E-05
0.94	0.0000	0.0045	0.0000	0.0000E+00
0.85	0.0000	0.0044	0.0000	0.0000E+00

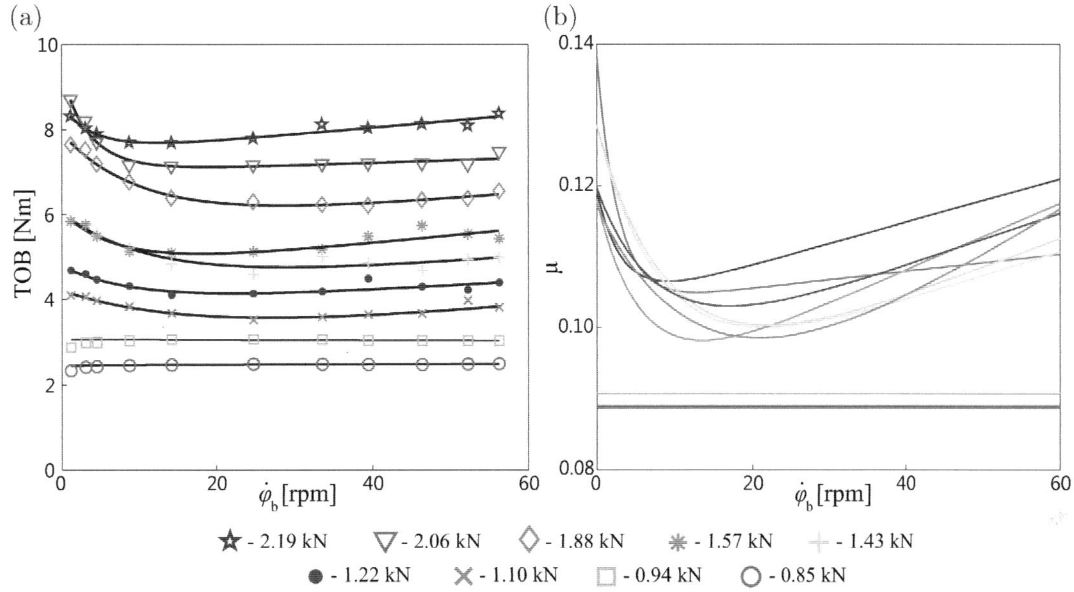

Figure 3. (a) Experimental identification of TOB curves for different WOB conditions (b) Corresponding equivalent friction coefficients as a function of bit velocity.

3. Experimental verification

Firstly we consider a response of stick-slip oscillations, for initial pre-buckling of $d_{pb} = 0.0381$ m (axial downward displacement) and WOB value of $W_b = 1.45$ kN, whose time history and phase portrait are shown in Fig. 4 (a). As can be seen, one observe stick-slip oscillations of constant amplitude, which are the result of small oscillations of the top speed and induced buckling of the flexible shaft. In order to replicate this kind of response in the FE model, we fit the top excitation (black curves in Figs. 4 (a)-(b)) using Fourier series:

$$\dot{\varphi}_t = a_0 + \sum_{n=1}^{6} a_n \cos(n\omega_p t) + \sum_{n=1}^{6} b_n \sin(n\omega_p t), \tag{4}$$

where $a_0 = 3.421\,\mathrm{rad/s}$, $a_1 = 0.0001082\,\mathrm{rad/s}$, $b_1 = 0.001583\,\mathrm{rad/s}$, $a_2 = 0.0006394\,\mathrm{rad/s}$, $b_2 = -0.001014\,\mathrm{rad/s}$, $a_3 = 0.0001515\,\mathrm{rad/s}$, $b_3 = -0.0009761\,\mathrm{rad/s}$, $a_4 = -0.0006725\,\mathrm{rad/s}$, $b_4 = 0.001115\,\mathrm{rad/s}$, $a_5 = -0.01963\,\mathrm{rad/s}$, $b_5 = -0.01733\,\mathrm{rad/s}$, $a_6 = 0.0001247\,\mathrm{rad/s}$,

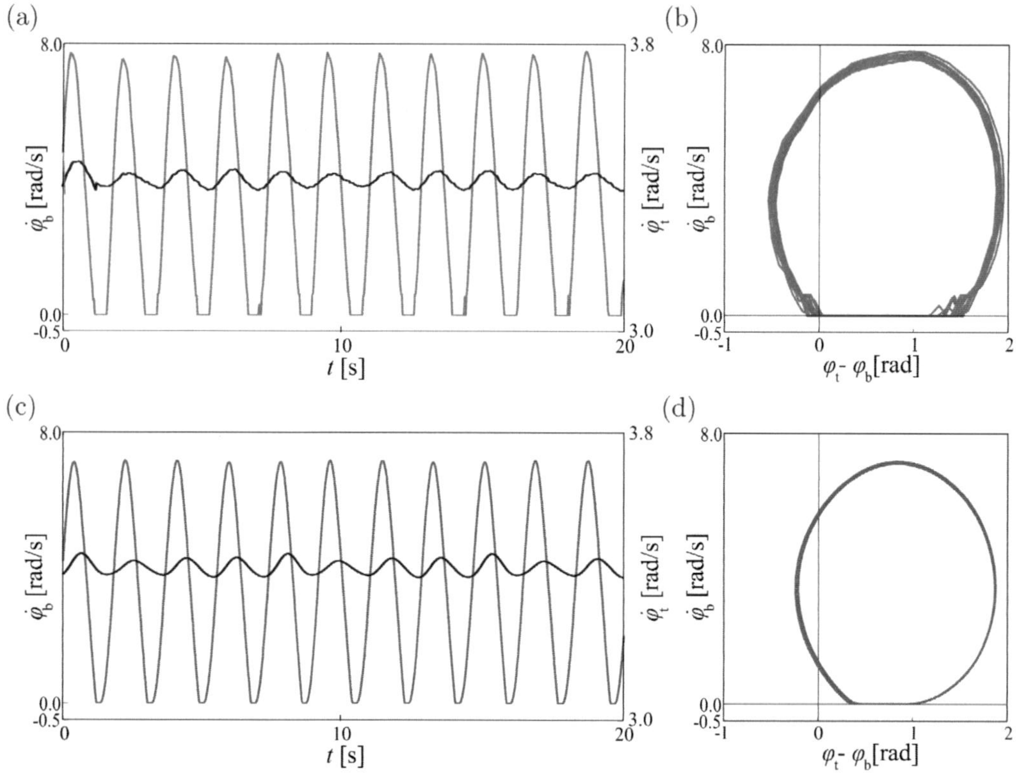

Figure 4. An example of stick-slip oscillations occurring in the experimental rig for pre-buckled configuration (d_{pb} = 0.0381 m) and W_b=1.45 kN (time history (a), phase portrait (b)) and corresponding response for FE model (time history (c), phase portrait (d))

$b_6 = -0.003006\,\mathrm{rad/s}$, $\omega_p = 0.6836\,\mathrm{rad/s}$. As a next step we choose the appropriate parameters for TOB model (see Eq. (3) and Table 1), that correspond to the measured WOB of $W_b = 1.45$ kN. As can be seen in Fig. 1 (b), we apply initial axial displacement $d_{pb} = 0.381$ m at the top of the flexible shaft, in order to induce buckling of the flexible shaft, which corresponds directly to the experimental conditions. For this particular case we assign corresponding mass and inertia to the reduced drill-bit and BHA (see Fig. 1 (b)): $m_{\mathrm{BHA}} = 148.579$ kg and $J_{\mathrm{BHA}} = 0.8416$ $\mathrm{kg\,m^2}$, taking into account effect of gravity ($g = 9.81$ m/s^2). The following parameters of the flexible shaft: $E_1 = E_2 = 3.64\,\mathrm{GPa}$, $E_3 = 1.50\,\mathrm{GPa}$, $G_{12} = 0.02\,\mathrm{GPa}$, $G_{13} = G_{23} = 18.16\,\mathrm{GPa}$, $\nu_{12} = \nu_{13} = \nu_{23} = 0.30$ and $\alpha = 7.50$ have been used. As can be seen in Fig. 1(b), the FE model of the experimental setup consists of two cylinders, one representing the drill-bit and the other the rock sample, and the flexible shaft set in the prebuckled configuration, which allows us to observe helical buckling condition.

As can be seen in Fig. 4(b), we are able to obtain a good qualitative and quantitative agreement between experiment and our FE model. Note, that there are small differences, as the model predicts slightly smaller amplitude of stick-slip vibrations, while keeping the same period. At the same time, when comparing the phase portraits for both cases, we observe that the sticking interval is also slightly smaller than the one observed in experiment, which might be caused by the simplification of cutting process using an equivalent friction approach.

In Fig. 6 (a)-(b) and Fig. 5 (a)-(b) we present the comparison of the shapes of the flexible shaft, during different stages of observed stick-slip oscillations, for front and top view respectively. Note, that the scale of pictures taken during experiment and from the FE model is not the

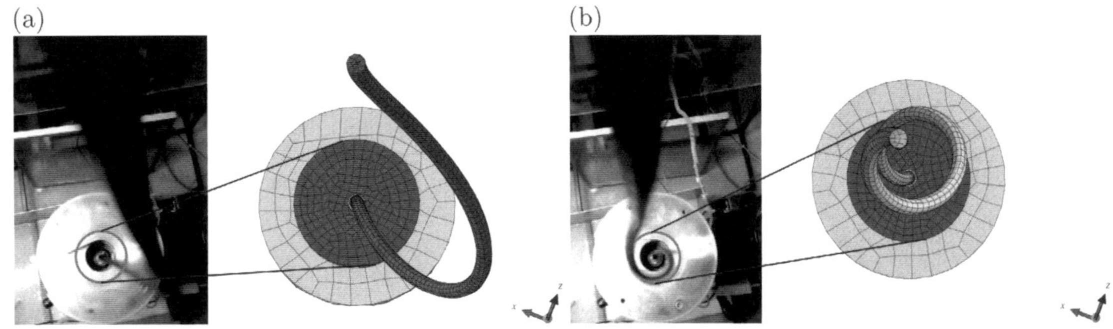

Figure 5. Comparison of shape of the flexible shaft during stick-slip oscillations for experiment and FE model (top view), whose time history is shown in Fig. 4

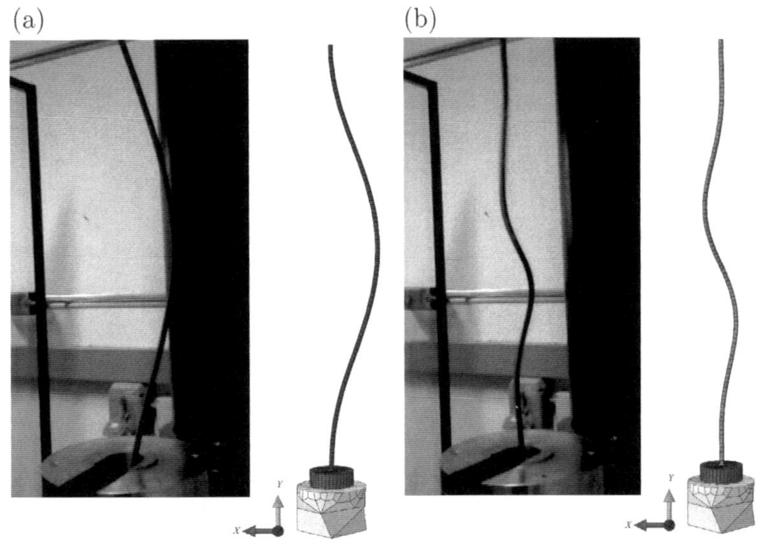

Figure 6. Comparison of shape of the flexible shaft during stick-slip oscillations for experiment and FE model (front view), whose time history is shown in Fig. 4

same. The size of the red circle, as introduced in photos from experiment(Fig. 5 (a)-(b)) corresponds directly to the size of blue cylinder representing the drill-bit in the FE model. As the drill-bit is in motion, the flexible shaft takes the shape of the initial configuration as shown in Fig. 6 (a) and Fig. 5 (a), but once the drill-bit stops we observe an abrupt change, as the flexible shaft takes the shape of a helix (see Fig. 6 and Fig. 5 (b)). As during the sticking phase, the top continues to rotate, we observe an increasing torsional deformation of the flexible shaft, until the moment that enough energy to overcome the static torque is stored in the shaft. Once that happens we observe unwinding of the helix formed. Finally the flexible shaft takes the shape of Fig. 6 (a) and Fig. 5 (a) and the whole cycle is repeated. Clearly, from Figs. 5 and 6 it is visible that we are able to obtain a good agreement between the experimental time histories and the FE predictions.

4. Conclusions and future work

In this paper, an experimental verification of the developed FE model of the drilling rig introduced in [7, 6] is carried out. In general a flexible shaft used to replicate the drill-string in the experimental conditions can be driven in two configurations: straight and buckled. In this work we concentrated our attention on latter case, to shed some light on the complexity of the helical buckling phenomenon. Thereby, based on a careful identification of all the experimental parameters, one is able to adjust the FE model to replicate appropriately the results recorded on the experimental rig. It is demonstrated that both the qualitative and quantitative agreements between experiments and the FE model are achievable. The developed model is capable of predicting accurately the shape of the flexible shaft during different phases of stick-slip phenomena, as shown in Figs. 5 and 6. In view of positive verification of the developed FE model, we can conclude that the model applicability has been demonstrated and further analysis of the model will be conducted, with a particular attention to parameters controlling onset of stick-slip phenomenon.

Acknowledgements

The authors wish to thank Dr. Joseph Páez Chávez for his help in experimental work and acknowledge the financial support of BG Group plc.

References

[1] H. Al-Naser and Y.A. Khulief. Finite element dynamic analysis of drillstrings. *Finite Elements in Analysis and Design*, 41(13):1270–1288, 2005.

[2] A.M Chevallier, N.P Politis, M.L Payne, and P.D. Spanos. Oil and gas well drilling: A vibrations perspective. *Shock and Vibration Digest*, 35(2):85–103, 2003.

[3] Dassault Systemes Simulia Corp. *ABAQUS Theory Manual and Analysis User's Manual (Version 6.12)*, 2010.

[4] H. Hakimi and S. Moradi. Drillstring vibration analysis using differential quadrature method. *Journal of Petroleum Science and Engineering*, 70(3-4):235–242, 2010.

[5] J. D. Jansen and L. van den Steen. Active damping of self-excited torsional vibrations in oil well drillstrings. *Journal of Sound and Vibration*, 179(4):647–668, 1995.

[6] M. Kapitaniak. *Nonlinear Dynamics of Drill-strings*. PhD thesis, University of Aberdeen, 2015.

[7] M. Kapitaniak, V. Vaziri Hamaneh, J. Páez Chávez, and M. Wiercigroch. Unveiling Complexity of Drill-String Vibrations: Experiments and Modelling. *International Journal of Mechanical Sciences*, 101-102:324–337, 2015.

[8] Al-Naser H. Khulief, Y.A. Finite element dynamic analysis of drillstrings. *Finite Elements in Analysis and Design*, 41(13):1270–1288, 2005.

[9] R. I.. Leine, D. H. van Campen, A. de Kraker, and L. van den Steen. Stick-Slip vibrations induced by alternate friction models. *Nonlinear Dynamics*, 16(1):41–54, 1998.

[10] C. Liao, B. Balachandran, M. Karkoub, and Y.L. Abdel-Magid. Drill-string dynamics: Reduced-order models and experimental studies. *Journal of Vibration and Acoustics*, 133((4)), 2011.

[11] E. M. Navarro-López. An alternative characterization of bit-sticking phenomena in a multi-degree-of-freedom controlled drillstring. *Nonlinear Anal. Real World Appl.*, 10(5):3162–3174, 2009.

[12] E. M. Navarro-López and D. Cortés. Avoiding harmful oscillations in a drillstring through dynamical analysis. *Journal of Sound and Vibration*, 307:152–171, 2007.

[13] J. Wojewoda, A. Stefanski, M. Wiercigroch, and T. Kapitaniak. Hysteretic effects of dry friction: modelling and experimental studies. *Philosophical Transactions of the Royal Society A*, 366:747–765, 2008.

Active vibration control using mechanical and electrical analogies

A Torres-Perez, A Hassan, S Kaczmarczyk, P Picton
The University of Northampton, St. George's Avenue, Northampton, NN2 6JD, United Kingdom
E-mail: Ali.hassan@northampton.ac.uk,

Keywords: active DVA, lumped electrical-mechanical circuit, voice coil motors

Abstract. Mechanical-electrical analogous circuit models are widely used in electromechanical system design as they represent the function of a coupled electrical and mechanical system using an equivalent electrical system. This research uses electrical circuits to establish a discussion of simple active vibration control principles using two scenarios: an active vibration isolation system and an active dynamic vibration absorber (DVA) using a voice coil motor (VCM) actuator. Active control laws such as gain scheduling are intuitively explained using circuit analysis techniques. Active vibration control approaches are typically constraint by electrical power requirements. The electrical analogous is a fast approach for specifying power requirements on the experimental test platform which is based on a vibration shaker that provides the based excitation required for the single Degree-of-Freedom (1DoF) vibration model under study.

1 Introduction

Dynamic Vibration Absorbers (DVAs) were patented by Herman Frahzm in 1909 [1] and the principles of passive DVAs design were fully described by Ormondroyd and Den Hartog [2, 3].

DVAs has been extensively studied in building structures, as defence mechanism against earthquakes, to counter seismic movements and wind forces [4]. DVAs are widely used to control vibration in structures. Passive control and active control are the typical methods for mitigating undesired vibrations. Active control requires sensors and actuators and electronic control systems to reduce vibration levels [5].

Coupled electrical and mechanical systems are typically analysed using equivalent circuit models. Equivalent circuit representations are widely used in electroacoustic [6] and electromechanical transducer analysis [7] due to their powerful visualization of the underlying physical phenomena. This study analyses an active DVA and an active vibration isolation system using electrical analogous circuit models.

2 Modelling the experimental active vibration control rig.

The experimental rig shown in Figure 1a. was built to test the concept of active control using an adjustable mass connected to a planar spring and one VCM actuator. Vibration measurements are performed using two accelerometers, one connected to the absorber mass and the other one to the shaker table. Two electrical signals are recorded for VCM characterization that correspond to the VCM current and voltage which are measured using one isolated differential voltage probe and one current probe respectively. The physical model which is a pictorial representation of the system under study is shown in Figure 1b.

The vibration model corresponds to a 1DoF with base excitation as shown in Figure 2.

The differential equation that describes the motion of the mass 'm' using 2nd Newton's law is shown in Equation 1.

$$m\ddot{x}_1 = U - k\left(x_1 - x_2\right) - c\left(\dot{x}_1 - \dot{x}_2\right) \tag{1}$$

where x_2 is the base excitation which is provided by the shaker table , x_1 describes the spring suspended mass motion, k is the helical spring stiffness, c is the mechanical damping and U is a generic actuator force.

Rig specifications are shown in Table 1. The VCM is a GVCM-051-051-01 manufactured by Moticont. VCM impedance measurements were taken using the frequency response analyser PSM1735 with an IAI LCR interface manufactured by Newtons4th.

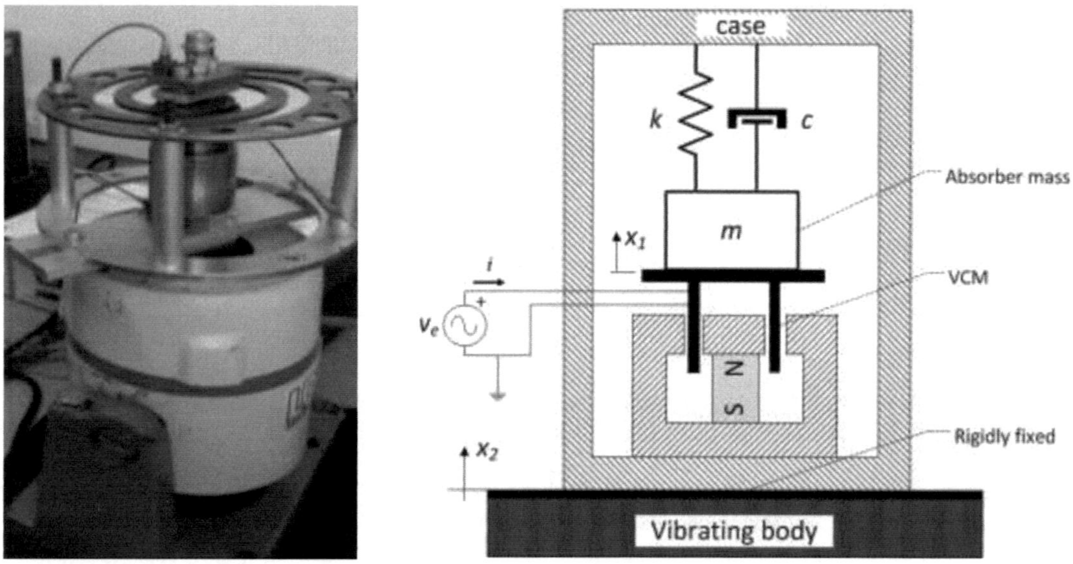

a) Experimental rig **b) Physical model**

Figure 1. Active vibration control rig

Table 1 Test rig parameters

Parameter	Value	Unit	Parameter	Value	Unit
m	432.4	[g]	$L_E\Big\|_{f=50Hz}^{\text{no magnet air coil}}$	2.1	[mH]
c	0.5 Negligible	[Ns/m]	$R_E\Big\|_{f=50Hz}^{\text{no magnet air coil}}$	3.08	[Ω]
k	42.167	[kN/m]	$L_E\Big\|_{f=50Hz}^{\text{with magnet}}$	4.1	[mH]
$f_0 = \frac{1}{2\pi}\sqrt{\frac{k}{m}}$	49.7	[Hz]	$R_E\Big\|_{f=50Hz}^{\text{with magnet}}$	3.3	[Ω]
			$\alpha_{VCM}\Big\|_{datasheet}$	8	[N/A]

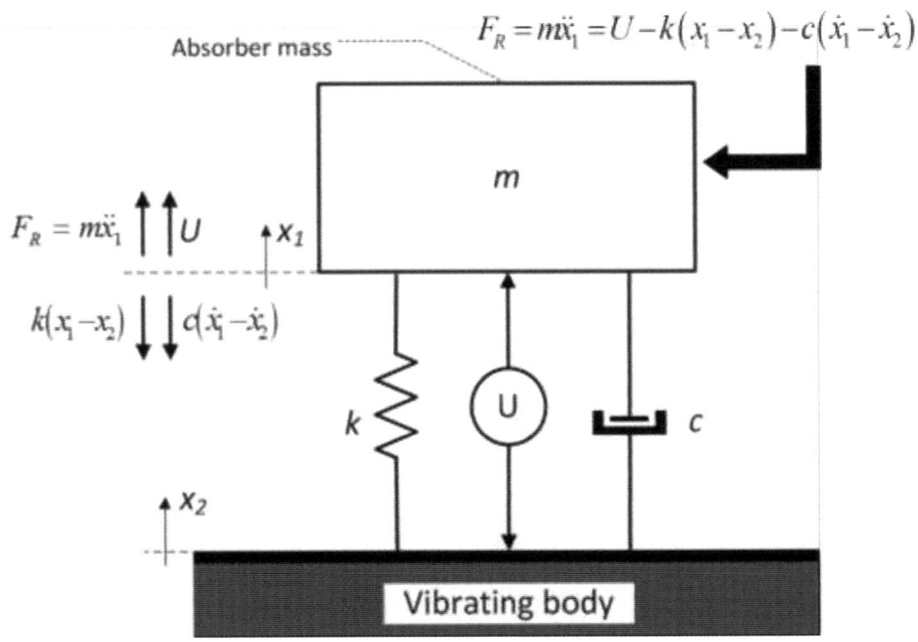

Figure 2. Active control vibration model with a generic actuator U (1DoF)

2.1 Equivalent circuit representation of the 1DoF active control rig with a generic actuator

An analogous electrical and mechanical system will have differential equations of the same form. There are two analogues for the system shown in Figure 2. Type I analogues use current as mechanical force representation and Type II analogues use voltage as force representation. Table 2 shows the relationship between electrical and mechanical quantities.

Table 2 Interpretation of electrical and mechanical quantities

Electrical Quantity	Mechanical Analog I (Force-Current)	Mechanical Analog II (Force Voltage)
Voltage, e	Velocity, v	Force, f
Current, i	Force, f	Velocity, v
Resistance, R	Lubricity, 1/B (Inverse friction)	Friction, B
Capacitance, C	Mass, M	Compliance, 1/K (Inverse spring constant)
Inductance, L	Compliance, 1/K (Inverse spring constant)	Mass, M

Type I is represented in Figure 3 whilst Type II in Figure 4. A circuit analysis using Kirchoff's Laws yields the same differential equation as shown in Equation 1. For Type I, the differential equations are obtained using KCL in node A as shown in Equation 2 and for Type II, they are obtained using KVL in loop I as shown in Equation 3.

$$\left(\text{KCL N}_\text{A}\right)\ C\frac{d\dot{x}_1}{dt}+\frac{\left(\dot{x}_1-\dot{x}_2\right)}{R}+\frac{1}{L}\int\left(\dot{x}_1-\dot{x}_2\right)dt=u$$

$$\left(\text{KCL N}_\text{A}\right)\ m\ddot{x}_1+c\left(\dot{x}_1-\dot{x}_2\right)+k\left(x_1-x_2\right)=u \tag{2}$$

initial conditions

$\text{U}\left(0\text{s}\right)=0\ ;\ \text{x}_1\left(0\text{s}\right)=\text{x}_1\left(0\text{s}\right)=0$

$$\left(\text{KVL Loop 1}\right)\ u+\frac{1}{C}\int\left(\dot{x}_2-\dot{x}_1\right)dt+R\left(\dot{x}_2-\dot{x}_1\right)-L\frac{d\dot{x}_1}{dt}=0$$

$$\left(\text{KVL Loop 1}\right)\ u+k\left(x_2-x_1\right)+c\left(\dot{x}_2-\dot{x}_1\right)-m\ddot{x}_1=0 \tag{3}$$

initial conditions

$\text{U}\left(0\text{s}\right)=0\ ;\ \text{x}_1\left(0\text{s}\right)=\text{x}_1\left(0\text{s}\right)=0$

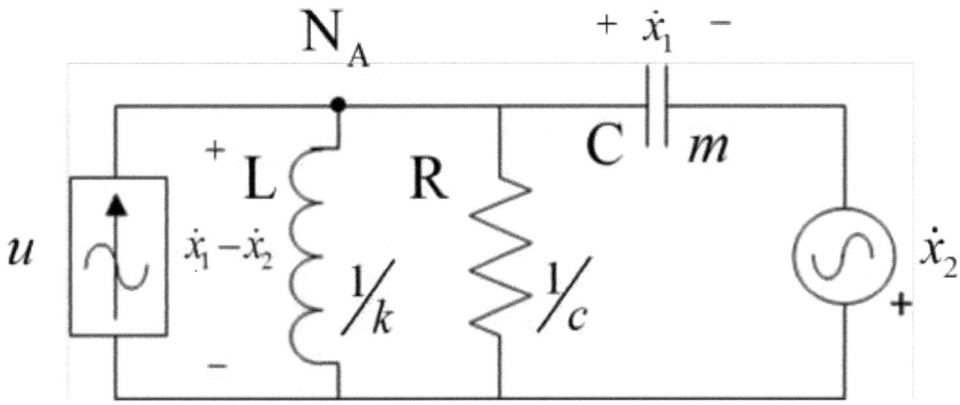

Figure 3. 1DoF active control rig with generic actuator (Type I)

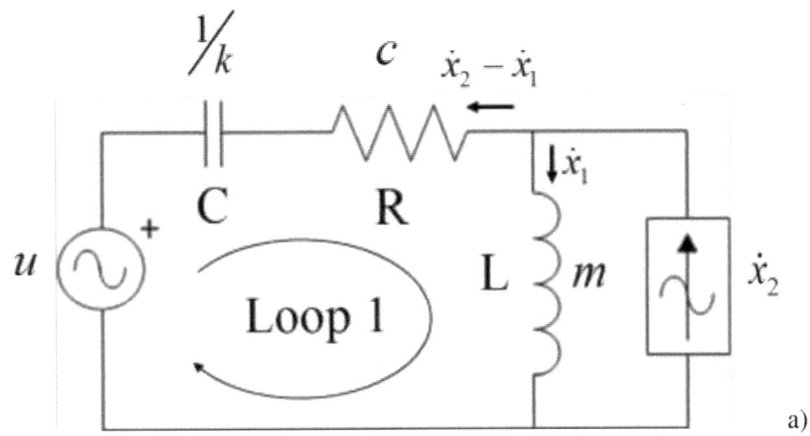

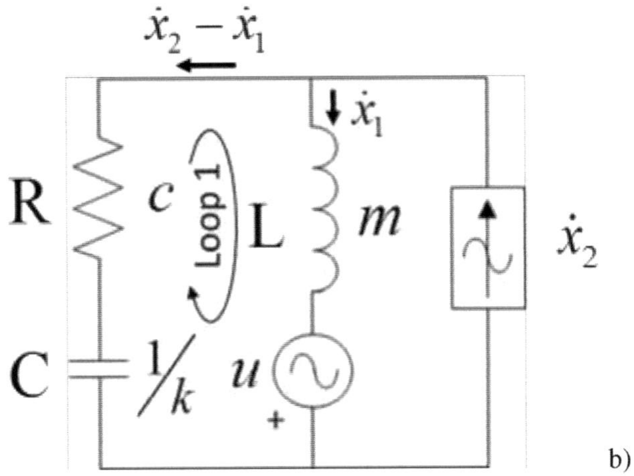

Figure 4. 1DoF active control rig with generic actuator (Type II)

2.2 *Active vibration control strategies for two real scenarios: vibration isolation and vibration absorbers.*

The concept of vibration isolation is that the mass 'm' is held motionless. That implies that the speed $\dot{x}_1$ must be 0. Using the circuit shown in Figure 4, vibration isolation implies that the current $\dot{x}_1$ must be 0. Using the simplified circuit in Figure 5 the control law could be derived by using superposition theorem and the current divider rule as shown in Equation 4. Electrical variables such as current and voltages are expressed as forces and velocities respectively.

$$(sX_1) = (sX_2)\frac{Z_T}{Z_{MP}} + \frac{U}{Z_{MS}+Z_{MP}} = (sX_2)\frac{Z_{MS}}{Z_{MS}+Z_{MP}} + \frac{U}{Z_{MS}+Z_{MP}} \tag{4}$$

If we want to minimize x1 motion → $\dot{x}_1 = 0$ then $U = -(sX_2)\cdot Z_{MS}$

$$Z_{MS}(s) = c + \frac{k}{s}$$

$$U(s) = -(sX_2)\cdot Z_{MS} \rightarrow U(s) = -(cs+k)X_2(s) \qquad (5)$$

The amplitude of an undamped DVA tends to be infinite at its resonant frequency. The optimum active control laws could be derived from the circuit in Figure 5. The actuator force u must be chosen for an infinite actuator stroke at the excitation frequency of $\dot{x}_2$. An active DVA that mimics a passive DVA without damping at any frequency implies an actuator speed or current $\dot{x}_1 - \dot{x}_2$ of infinite value. This condition is easily obtained using Kirchoff's Voltage Laws after applying the source transformation theorem as shown in Figure 6a. and Equation 6.

$$\dot{x}_1 - \dot{x}_2 = \infty \text{ only if } U + s(X_2 - X_1)Z_{MS} + s(X_2 - X_1)Z_{MP} = 0$$

$$U(s) = -(X_2(s) - X_1(s))(ms^2 + cs + k), \text{ if } s = j\omega \qquad (6)$$

$$U(j\omega) = -(X_2(j\omega) - X_1(j\omega))(c + j\omega(m-k))\cdot(j\omega) = -(X_2(j\omega) - X_1(j\omega))\left(c + j\left(\omega m - \frac{k}{\omega}\right)\right)\cdot(j\omega)$$

If we use the circuit in Figure 6b. the condition is achieved in Equation 7.

$$\dot{x}_1 = \infty \text{ only if } -U + (sX_1)\cdot(Z_{MS} + Z_{MP}) = 0$$

$$U(s) = X_1(s)(ms^2 + cs + k) \qquad (7)$$

$$U(j\omega) = X_1(j\omega)\cdot(c + j\omega(m-k))\cdot(j\omega) = X_1(j\omega)\left(c + j\left(\omega m - \frac{k}{\omega}\right)\right)\cdot(j\omega)$$

According to Equation.6 and Equation 7, the actuator force must remove the damping of the system and it should provide a force proportional to actuator relative position $(x_1 - x_2)$ of magnitude α or acceleration $(\ddot{x}_1 - \ddot{x}_2)$ of magnitude γ. Gain scheduling control consists on two possible gains, one proportional to acceleration (α) and the other proportional to position (γ) are defined in Equation 7 using the natural resonant frequency of the absorber mass [8].

$$U(j\omega) = -(X_2(j\omega) - X_1(j\omega))\left(c + j\omega\underbrace{\left(m - \frac{k}{\omega^2}\right)}_{\alpha}\right)\cdot(j\omega) = -(X_2(j\omega) - X_1(j\omega))\left(c + \frac{j}{\omega}\underbrace{(\omega^2 m - k)}_{\gamma}\right)\cdot(j\omega)$$

Using the natural resonant frequency $\omega_0 = \sqrt{\frac{k}{m}}$ (9)

(Acceleration gain) $\alpha = \left(m - \frac{k}{\omega^2}\right) \xrightarrow{\omega_0 = \sqrt{\frac{k}{m}}} \alpha = m\left(1 - \left(\frac{\omega_0}{\omega}\right)^2\right)$

(Position gain) $\gamma = (\omega^2 m - k) \xrightarrow{\omega_0 = \sqrt{\frac{k}{m}}} \gamma = m(\omega^2 - \omega_0^2)$

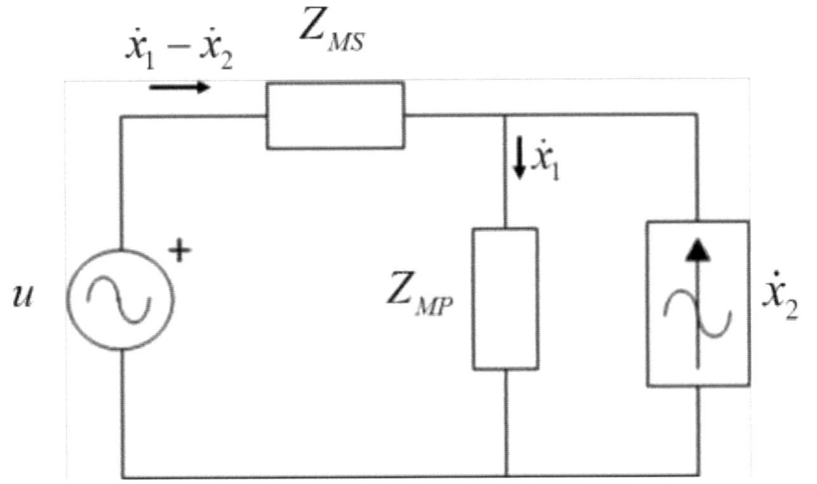

$$Z_{MS}(s) = R + Z_C = c + \frac{k}{s}$$

$$Z_{MP}(s) = ms$$

$$Z_T(s) = Z_{MS} \parallel Z_{MP} = \frac{Z_{MS} \cdot Z_{MP}}{Z_{MS} + Z_{MP}}$$

Vibration isolation $\dot{x}_1 = 0$

Figure 5. Actuator force for active vibration isolation

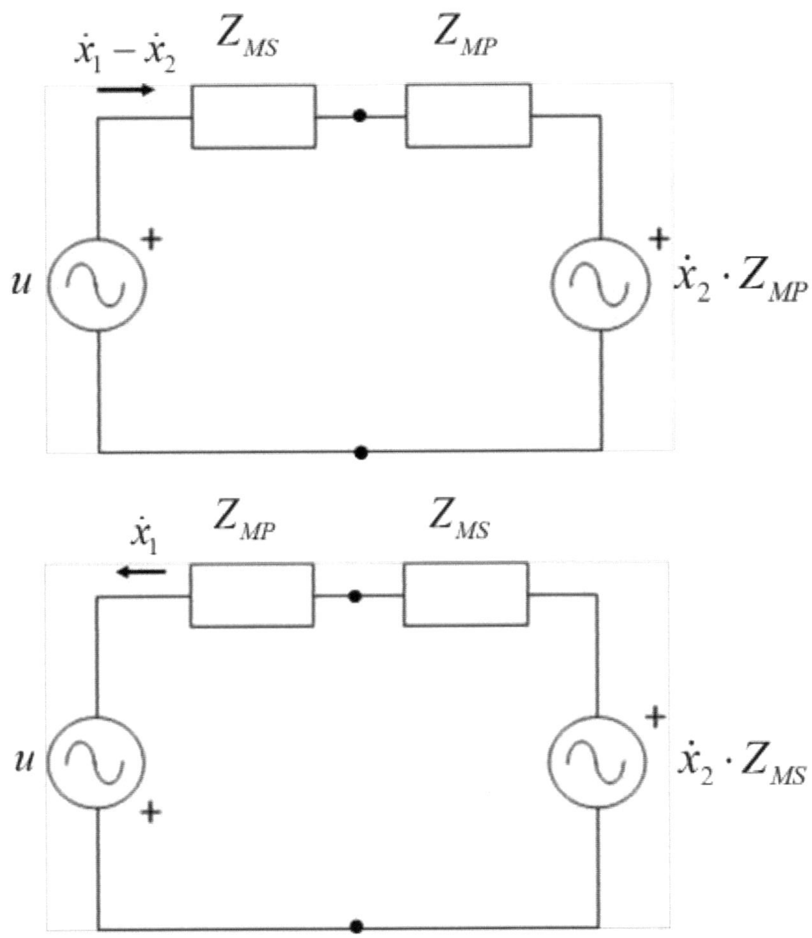

Active DVA.

Using circuit in Fig 4a

$\dot{x}_1 - \dot{x}_2 = \infty$ at resonance when $\dot{x}_2 \neq 0$

a)

Active DVA

Using circuit in Fig 4b

$\dot{x}_1 = \infty$ at resonance when $\dot{x}_2 \neq 0$

b)

Figure 6. Actuator force for an Active DVA

2.3 *Equivalent circuit representation of the 1DoF active control rig with a VCM*

If the chosen actuator is a Voice Coil Motor, the coupling equation between the electrical and mechanical system is due to Lorentz Force and Faradays's Law of Induction. The actuator mechanical force is proportional to actuator's current if there is a constant magnetic flux density in the VCM gap for all possible actuator stroke positions $N\frac{d\Phi}{dx} = cte$. The Back Electromotive Force is proportional to the total number of turns and actuator speed as shown in Equation 9.

α_{VCM} is the motor constant

$\alpha_{VCM} = N\frac{d\Phi}{dx} = Bl$ where:

N is the number of turns,

B the magnetic flux density in the gap (9)

l, the length of the wounded wire

(Actuator mechanical force) $F = Bl \cdot I = \alpha_{VCM} \cdot l$

(Induced voltage BEMF) $E = -N\frac{d\Phi}{dt} = -N\frac{d\Phi}{dx}\frac{dx}{dt} = -\alpha_{VCM}\dot{x}$

A possible circuit representation that couples the mechanical system and electrical side uses a gyrator and it is shown in Figure 7. The governing differential equations for this circuit are shown in Equation 10.

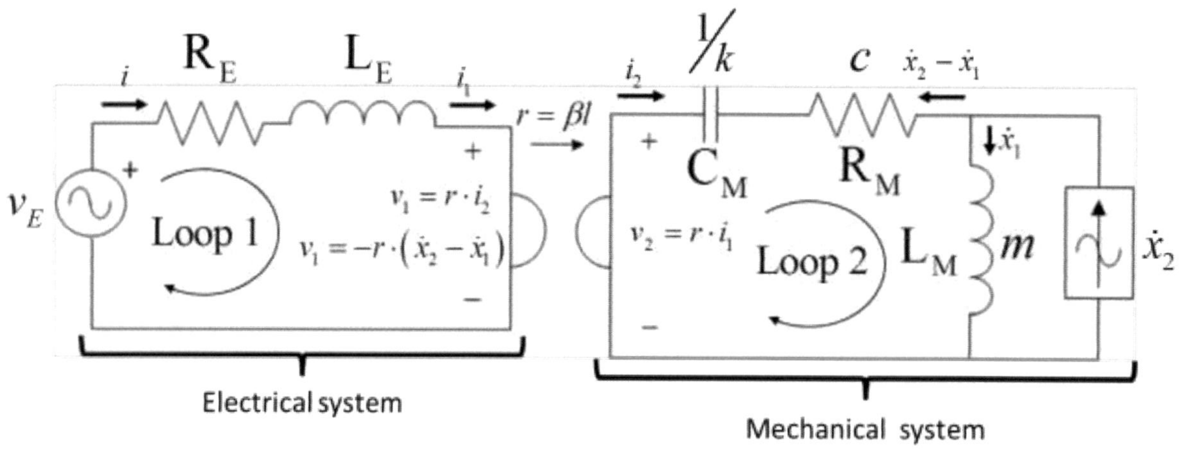

Figure 7. 1DoF active control rig with VCM (Type II)

$$(\text{KVL Loop 1})\ V_E - R_E \cdot i - L_E \cdot \frac{di}{dt} - v_1 = 0 \qquad (10)$$

$$\text{(KVL Loop 2)}\ v_2 + k(x_2 - x_1) + c(\dot{x}_2 - \dot{x}_1) - m\ddot{x}_1 = 0$$

$$r = \beta l = \alpha_{VCM} \quad \begin{cases} V_E - R_E \cdot i - L_E \cdot \dfrac{di}{dt} + r \cdot (\dot{x}_2 - \dot{x}_1) = 0 \\ r \cdot i_1 + k(x_2 - x_1) + c(\dot{x}_2 - \dot{x}_1) - m\ddot{x}_1 = 0 \end{cases}$$

For active vibration isolation or active control of a DVA, the VCM current should be same as the control laws shown in Equations 5-7 respectively. For example, ideal vibration isolation requires a VCM current according to Equation 11. As it can be seen, this control strategy only requires reading the shaker table position, speed or acceleration ($x_2, \dot{x}_2, \ddot{x}_2$). External electronics such as linear servo motor drives could adjust the VCM voltage to ensure the VCM current follows the desired control law.

$$u = \alpha_{VCM} \cdot i = \dot{x}_2 \cdot Z_{MS} \rightarrow I(s) = (sX_2) \cdot \frac{Z_{MS}}{\alpha_{VCM}} \rightarrow I(s) = \left(\frac{c}{\alpha_{VCM}} s + \frac{k}{\alpha_{VCM}} \right) X_2(s) \qquad (11)$$

2.4 *Electrical impedance and power requirements*

Electrical power requirements for active vibration control are one of the most important factors when choosing between Passive and Active solutions for mitigating vibrations in an industrial application. The circuit shown in Figure 8 corresponds to the equivalent impedance seen by the electrical system due to the mechanical one after impedance transformations.

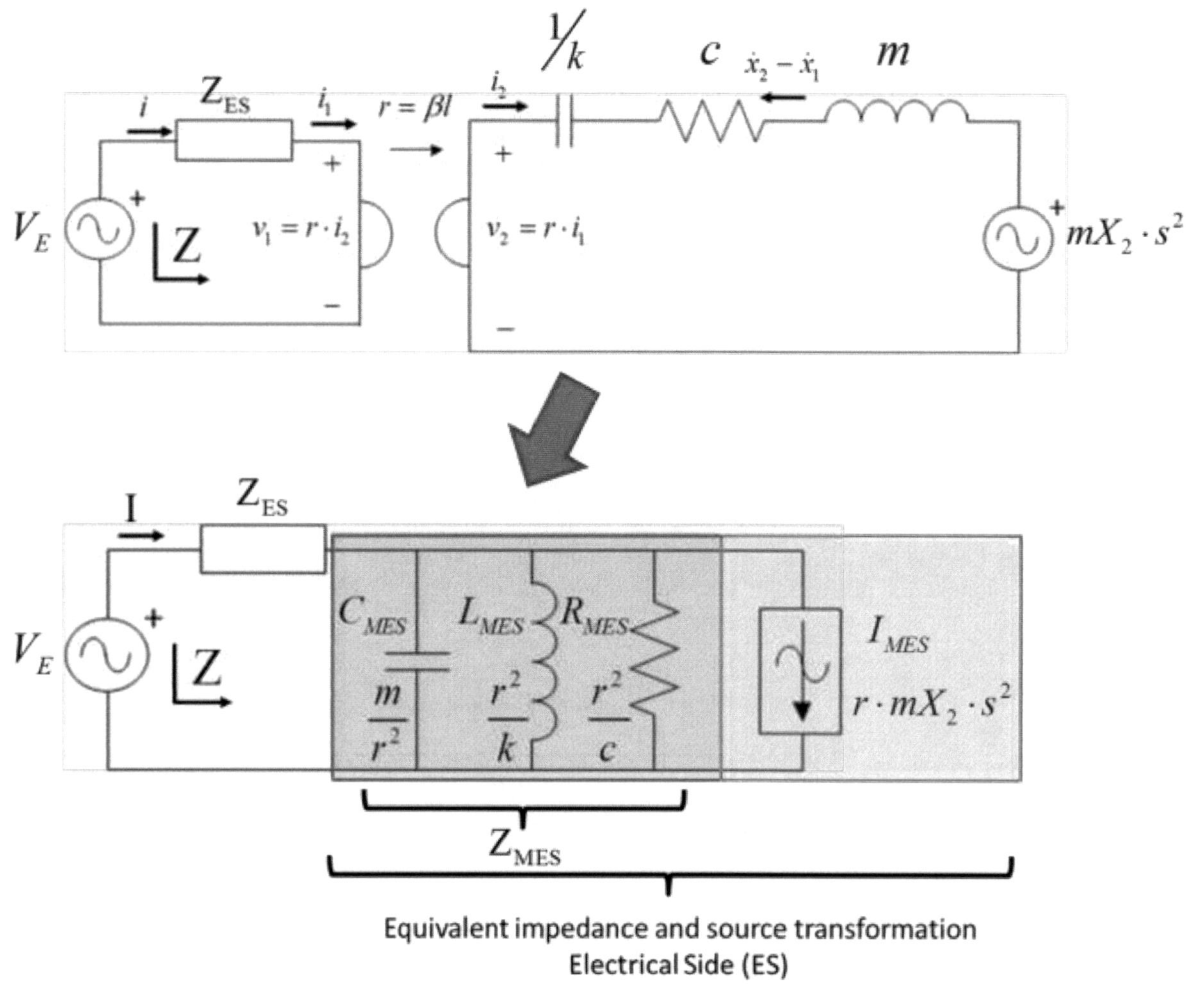

Figure 8. VCM Electrical impedance (Z)

An analysis of Figure 8 using source transformation theorem yields to the following electrical impedance expression as shown in Equation 12.

$$Z(s) = \frac{V_E(s) + I_{MES}(s)\ Z_{MES}(s)}{Z_{ES}(s) + Z_{MES}(s)}$$

where

$$Z_{MES}(s) = \frac{1}{\frac{1}{R_{MES}} + \frac{1}{s \cdot L_{MES}} + \frac{1}{1/(s \cdot C_{MES})}} \text{ and } Z_{ES}(s) = R_E + s \cdot L_E \tag{12}$$

If the shaker table does not move, the impedance equation could be further simplified as shown in Equation 13.

$$Z(s) = \frac{V_E(s)}{Z_{ES}(s) + Z_{MES}(s)} \quad \text{when} \quad X_2(s) = 0 \tag{13}$$

Electrical power requirements could be calculated using the impedance expressions as shown in Equation 14.

$$P_{ES}(s) = Z(s) \cdot I^2(s) = \frac{V_E^2(s)}{Z(s)} \begin{cases} \mathrm{Re}\{P_{ES}(s)\} \rightarrow \text{Active power} \\ \mathrm{Im}\{P_{ES}(s)\} \rightarrow \text{Reactive power} \end{cases} \tag{14}$$

2.5 *Mass on the spring motion using the equivalent circuit*

Using the circuit transformation shown in Figure 9 an equivalent circuit as shown in Figure 10 could be analysed for obtaining the mass motion $x_1, \dot{x}_1, \ddot{x}_1$

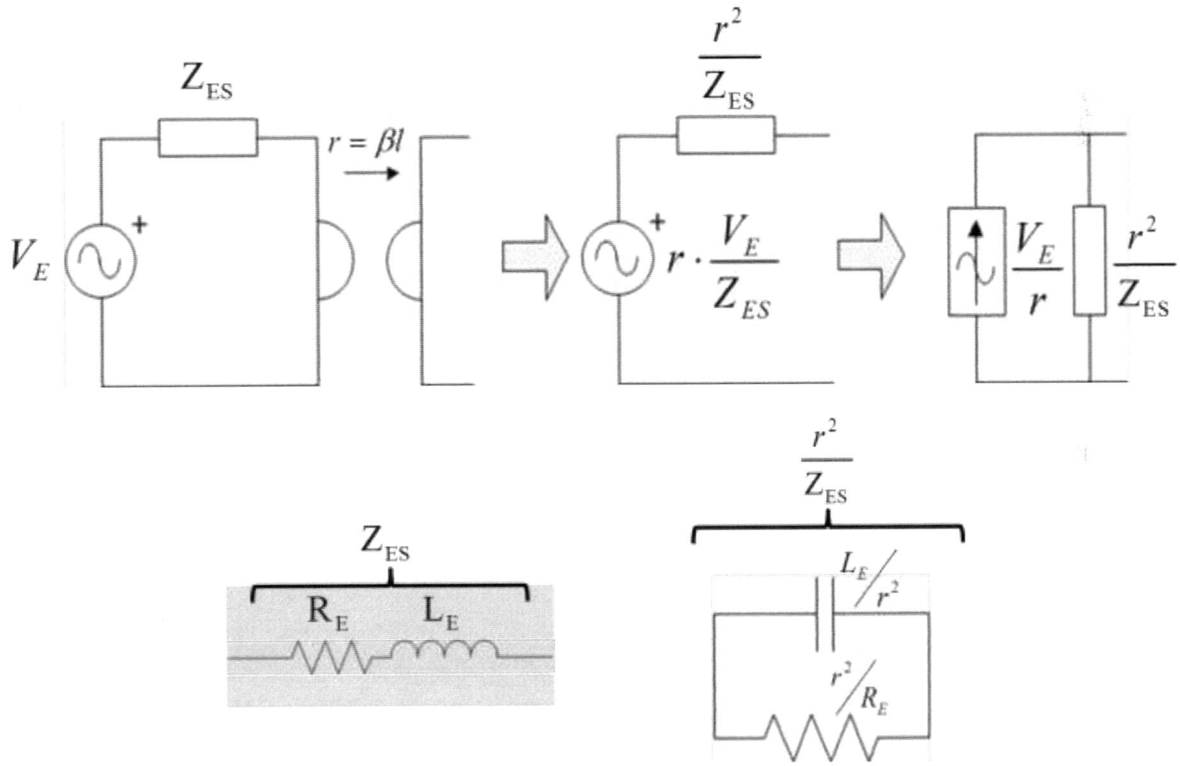

Figure 9. Equivalent circuit transformation to the Mechanical System

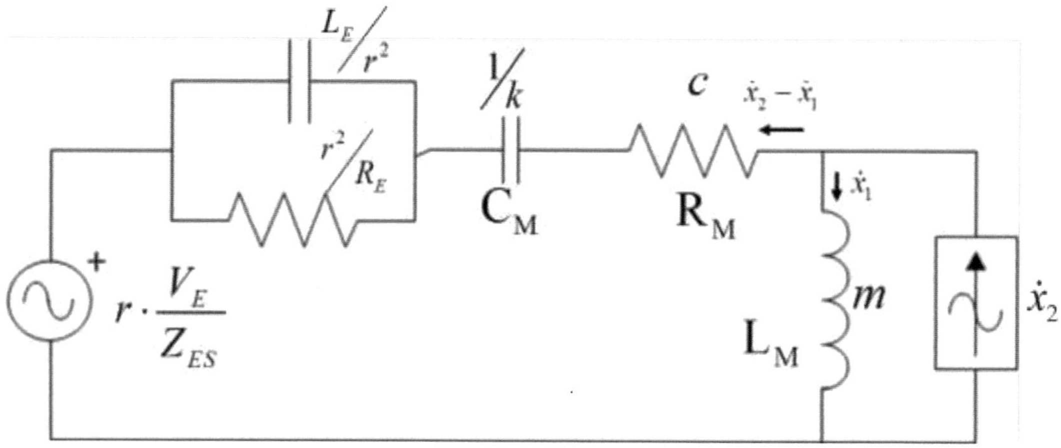

Figure 10. Equivalent circuit for absorber motion determination

The absorber mass motion could be analysed using Superposition Theorem and the current divider rule according to Equation 15.

$$\dot{x}_1 = \dot{x}_2 \frac{Z_T}{Z_{L_M}} + \frac{r \cdot \frac{V_E}{Z_{ES}}}{\frac{r^2}{Z_{ES}} + R_M + Z_{C_M} + Z_{L_M}} \quad \text{where}$$

$$Z_T = \frac{1}{\frac{r^2}{Z_{L_M}} + \frac{1}{\frac{1}{Z_{ES}} + R_M + Z_{C_M} + Z_{L_M}}} \qquad (15)$$

$$Z_{ES}(s) = R_E + s \cdot L_E \; ; R_M = c \; ; \; Z_{LM} = ms \; ; Z_{CM} = \frac{k}{s}$$

The circuit shown in Figure 10 describes the shift in resonant frequency when a short circuit is placed across the VCM terminals.

3 Model validation

Transmissibility experiments were carried out with the VCM terminals in open and closed circuit respectively. The model in Figure 7 could consider the added damping due to eddy current losses in the VCM actuator when there is an open circuit across the VCM terminals as show in Figure 11.

Fitting transmissibility plots to experimental results using the theoretical model in Equation 15 shows that the best damping factor is 6.1 N/(m/s) when the test was performed with an open circuit across VCM terminals and the damping factor is 20.2 N/(m/s) when a short circuit is placed across VCM terminals, Figures 12-13. Under short circuit test the resonant frequency shifted from 49.7Hz to 51Hz. This shift in frequency is explained in Figure10 as additional stiffness added by the VCM inductance (L_E).

$$\frac{X_2}{X_1} = \frac{(ms^2 + cs + k)}{(cs + k)} \tag{16}$$

The open circuit resistor value in this case is calculated according to Equation 17 if r is 8 N/A and $c|_{oc} = 6.1$ Ns/m and c=0.5Ns/m, therefore R_{OC} is 11.4Ω.

$$R_{OC} = \frac{r^2}{c|_{oc} - c} \tag{17}$$

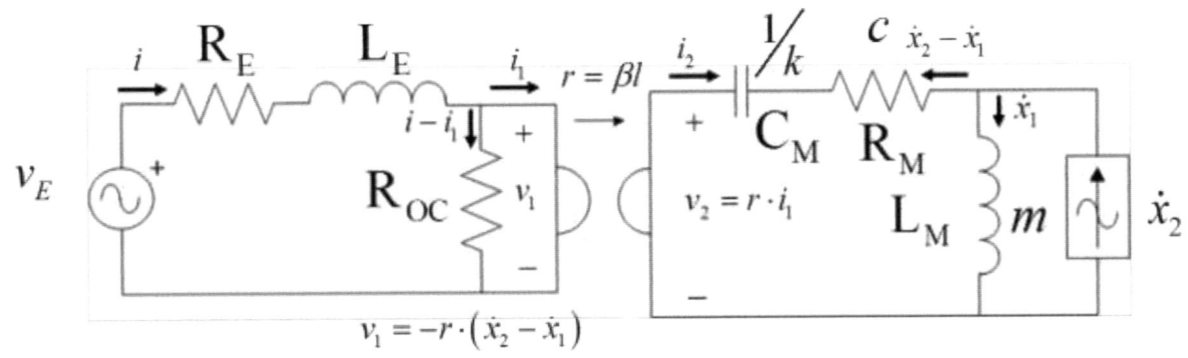

Figure 11. Model improvements to include VCM Open Circuit damping

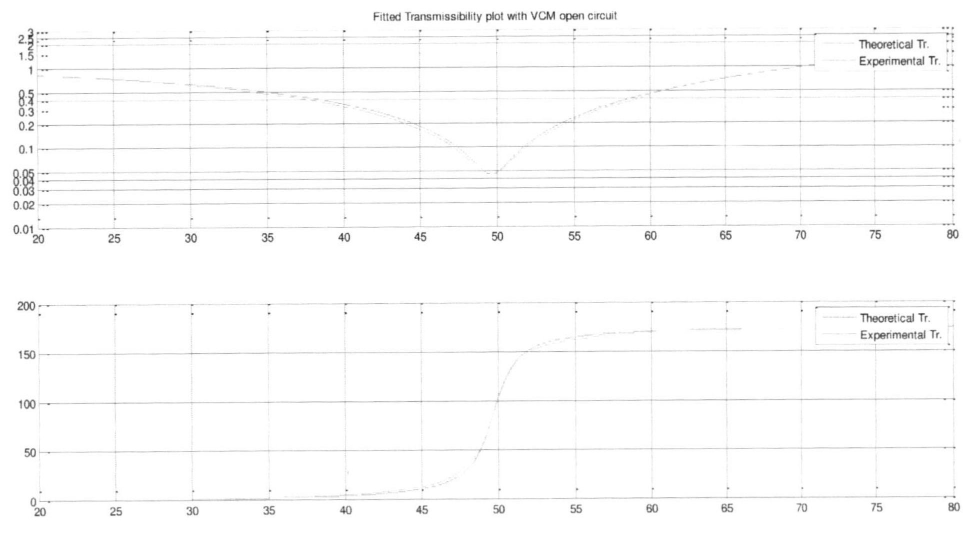

Figure 12. Transmissibility plots with an open circuit across VCM terminals

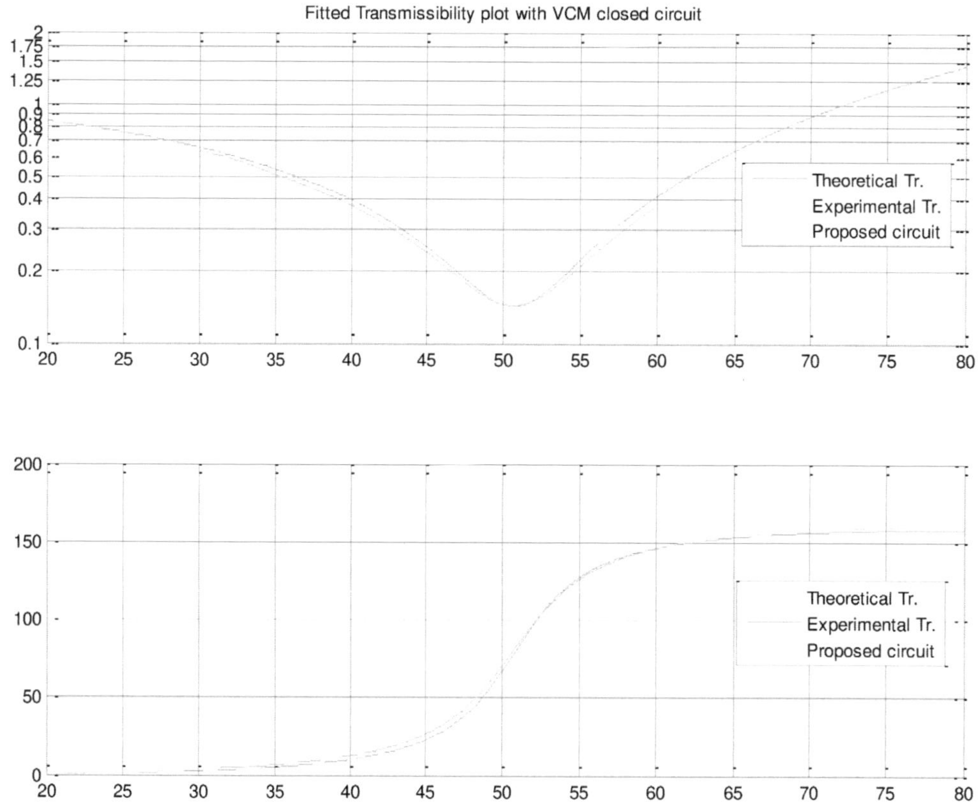

Figure 13. Transmissibility plots with a closed circuit across VCM terminals

The circuit shown in Figure 11 provides a trend that is in good agreement with experimental results when the VCM terminals are under open and closed circuit conditions respectively. The green trace in Figure 13 corresponds to the circuit values shown in Table 3.

Table 3 Circuit values

Parameter	Value	Unit	Parameter	Value	Unit
	432.4	[g]	L_E	4.1	[mH]
	0.5 Negligible	[Ns/m]	R_E	3.3	[Ω]
m	42.167	[kN/m]	$\alpha_{VCM} = r$	7.4	[N/A]
c	49.7	[Hz]	R_{OC}	9.78	[Ω]
k			C_M	23.71	[μF]
$f_0 = \frac{1}{2\pi}\sqrt{\frac{k}{m}}$			L_M	0.4324	[H]
			R_M	0.5	[Ω]

4 Conclusion

Coupled mechanical-electrical systems could be analysed with electromechanical analogous circuit models. The practicality of this method is that it provides a better visualization and interpretation of the system. A complete derivation of active control laws for dynamic vibration absorbers and vibration isolation is straightforward by analysing these circuits. Electrical power requirements could be defined in the frequency domain by calculating the electrical impedance seen by the input voltage source connected to the VCM. The proposed equivalent circuits show a good agreement with experimental results.

References

[1] Frahm, H., *Device for damping vibrations of bodies*, U.S. Patent No. 989958, 1909.

[2] Ormondroyd, J. and Den Hartog, J.P., *Theory of the dynamic vibration absorber,*

Transactions of the American Society of Mechanical Engineers, 50 (1928) , pp. 9–22.

[3] Den Hartog, J.P., *Mechanical Vibrations*, McGraw-Hill, New York, 1934

[4] Samali, B., Al-Dawod, M., Naghdy, F. *Active Control of Cross Wind Response of 76-Story Tall Building Using a Fuzzy Controller*, Journal of Engineering Mechanics, 130 (4) (2014), pp. 492-498.

[5] S. S. Rao, *Mechanical Vibrations.* 5th Edition. Pearson (2011).

[6] Marshall Leach, W. *On the specification of Moving-Coil Drivers for Low-frequency Horn-Loaded Loudspeakers*, Journal of the Audio Engineering Society 27(12) (1979), pp. 950-959.

[7] Tilmans, Harrie A.C. *Equivalent circuit representation of electromechanical transducers: I. Lumped-parameter systems*, Journal of Micromechanical Microengineering 6 (1996), pp. 157-176.

[8] R.F.A. Marques, D.A. Rade, S.S. Cunha, Jr., *Assessment of an Active Dynamic Vibration Absorber*, IMAC XIX - 19th International Modal Analysis Conference (2001), pp, 32-36.

Influences on lifetime of wire ropes in traction lifts

W Vogel [1 2]

[1] Institute of Mechanical Handling, University of Chemnitz, Germany

[2] Publicly certified expert for "Elevator-, Rope- and Lifting Technologies"

by IHK Region Stuttgart, Germany

Danziger Str. 46 D-73614 Schorndorf
E-mail: Wolfram_vogel@web.de

Abstract. Traction lifts are complex systems with rotating and translating moving masses, springs and dampers and several system inputs from the lifts and the users. The wire ropes are essential mechanical elements. The mechanical properties of the ropes in use depend on the rope construction, the load situation, nonlinearities and the lift dimensions. The mechanical properties are important for the proper use in lifts and the ride quality. But first of all the wire ropes (for all other suspension means as well) have to satisfy the safety relevant requirements sufficient lifetime, reliable determination of discard and sufficient and limited traction capacity. The lifetime of the wire ropes better the number of trips until rope discard depends on a lot of parameters of the rope and the rope application eg use of plastic deflection sheaves and reverse bending layouts. New challenges for rope lifetime are resulting from the more or less open D/d-ratio limits possible by certificates concerning the examination of conformity by notified bodies. This paper will highlight the basics of wire rope technology, the endurance and lifetime of wire ropes running over sheaves, and the different influences from the ropes and more and more important from the lift application parameters. Very often underestimated are the influences of transport, storage, installation and maintenance. With this background we will lead over to the calculation methods of wire rope lifetime considering the actual findings of wire rope endurance research. We´ll show in this paper new and innovative facts as the influence of rope length and size factor in the lifetime formular, the reduction of lifetime caused by traction grooves, the new model for the calculation in reverse bending operations and the statistically firmed possibilities for machine roomless lifts (MRL) under very small bending conditions.

1. The lifetime of a running lift rope

The ropes are central mechanical elements in traction lifts. The cabin and the counterweight, both connected to the building by guidance, are joined one with another by means of ropes and are connected to the drive sheave by the way of friction. The difference force resulting from the forces produced by the masses of cabin and counterweight will have to be transferred to the rope by the possible friction of the sheave groove geometry and the deflection angle.

When running over sheaves the suspension means are subjected to fluctuating bending and tension stresses, pressures and those stresses tend to force the suspension means to assume an oval cross-section. This is in sum a complex stress collective. The wear between the wires or the fibres in service has made it impossible until now to calculate the life time of the suspension means even where exact information is available about all stresses in the wires or fibres. The lifetime and the point of discard of the suspension means has to be established in series of bending fatigue tests. The statistical analysis of the bending fatigue tests lead to special lifetime formula depending of the test parameters. In real lift service the conditions will normally deviate from the test conditions. In preparation for forecasting lifetime for a lift which is being planned or has been built, the influence of the system parameters on the suspension means will have to be known; they will have to be taken account of by employing adjusted correction factors when calculating the number of trips.

2. Feyrer´s lifetime formula for steel wire ropes

The life time of a running rope is dependant upon rope parameters as rope design, rope diameter d, rope core, nominal wire strength, lubrication, etc. and upon parameters of the system traction lift such as diameter ratio sheave to rope D/d, tension force S, bending length, deflection angle, incline angle, groove profile, type of bending, etc.. Bending fatigue machines are in use to do the endurance tests on running ropes. The bending machines and the test procedure are described in [2].

The lifetime of steel wire ropes running over sheaves, that is the number of bending cycles until breakage under defined test conditions found in the tests, can be calculated for wire stress levels less than the so called Donandt force, i.e. before the yield stress of the wire is exceeded, using Feyrer´s first formular

$$lg\,\overline{N} = b_0 + \left(b_1 + b_4\,lg\frac{D}{d}\right)\left(lg\frac{Sd_0^2}{d^2 S_0} - 0{,}4\,lg\frac{R_0}{1770}\right) + b_2\,lg\frac{D}{d} + b_3\,lg\frac{d}{d_0} + \frac{1}{b_5 + lg\dfrac{l}{d}} \tag{1}$$

Feyrer´s formula was found by regression analysis. The parameters rope tension force S, nominal wire strength R_0, diameter ration sheave to rope D/d and bending length l are included.

Actually there is another lifetime formular very similar to eq (1) but with subticle destintion was created following developments of eg. rope industry. There is a scale factor for small rope diameter and the bending length. The performance in bending cycles is higher than expected with eq. (1). The new formula is

$$lg\,\overline{N} = b_0 + \left(b_1 + b_4\,lg\frac{D}{d}\right)\left(lg\frac{Sd_0^2}{d^2 S_0} - 0{,}4\,lg\frac{R_0}{1770}\right) + b_2\,lg\frac{D}{d} + lg\,f_d + lg\,f_L + lg\,f_C \tag{2}$$

The components f_d, f_L and f_C are discribing the scale influences of bending lenght and rope diameter. In detail the influence of bending length is

$$lg\,f_\ell = \frac{1{,}54}{2{,}54 + \left(\dfrac{\ell/d - 2{,}5}{57{,}5}\right)^{-0{,}14}} \tag{3}$$

and the influence of the rope diameter is

$$lg\,f_d = \frac{0{,}52}{-0{,}48 + \left(\dfrac{d}{16}\right)^{0{,}3}} \tag{4}$$

The regression coefficients in eq. (1) and eq. (2) have been calculated for the often used and the standardized rope designs on the basis of a lot of bending tests with different ropes of different manufacturers. We differ between the number of bending cycles until breakage and the number of bending cycles until discard. The individual results deviate around the calculated mean values. The test results must be statistically limited in N_{10} and N_{A10}. This means that with a safety of 95% only 10% of the ropes break earlier (N_{10}) or reach earlier the discard point (N_{A10}). The corresponding coefficients are summarized in [2].

3. Calculation of rope lifetime as per EN 81-1 and EN81-20/50, Annex N

Why two different formulars here? The reason is simple and explainable based on the works actualizing european lift standards. The working groups reveising EN81-1 finished their work before the new formular was common. But there is no safety problem and further works of the working groups with an additional ammendment for the EN81-50 are possible.

Actually both standards EN81-1 and EN81-20/50 are valid. This allows a dimensioning following both standards. The situation is tricky because both standards are covered in the moment by lift directive 95/16/EU until 19.04.2016. The the lift directive 2014/33/EU is then valid, only EN81-20/50 is valid and very important for manufacturer: without any time gap.

The method used to calculate the rope lifetime described in [2] is based on the statistical analysis of the bending fatigue tests, along with the consideration of the stresses in service and particular rope drive parameter which influence the rope lifetime. With detailed requirements for the minimum number of trips (3 years at Z=200.000 in one direction, [3]) this calculation method for steel wire ropes is used for dimensioning the rope drive concerning to the diameter ration sheave to rope D/d and the safety factor S_f in annex N, [1] and [4].

The most stressed rope zone is identified within the rope drive. This is the zone which experiences the most bends during a trip. The influence of the drive sheave on rope lifetime will be expressed as an equivalent number of sheaves with round groove. The equivalent number of sheaves for the rope drive sheave is determined by the type of groove (V-groove, round groove with undercut) and the geometric parameters for the groove. The correction factor from [2] are converted without change to achieve the equivalent number of pulleys $N_{equiv(t)}$. When determine the equivalent number of sheaves it is necessary to take into account the diameters of sheaves and the type of bending (simple bends N_{ps}, reverse bends N_{pr}). The damage potential of reverse bends is assigned by a factor of 4. Differing diameters at the drive sheave and the deflection sheaves are taken into account by the factor K_p. The safety factor S_f can be calculated using the equivalent number of sheaves N_{equiv}, the traction sheave diameter D_t and the rope diameter d_r. The basis are the regression coefficients from [2] and [4] for the statistical limited number of bendings until dicard for a rope with 6 parallel lay strands, 19 wires per strand and a fibre core.

In EN81-1, Annex N and EN81-20/50, Annex N the equivalent number of sheaves are equal in the case of grooves with undercut. The equivalent number of sheaves N_{equi} of the grooves with V-profile are in the future higher. The reasons are new results of research works [5] and longterm experiences of the lift manufacturers.

4. Reducing the diameter ratio D/d

Against the background of cost reduction on the drive side the trend in lift industry is to reduce the sheave diameter either by smaller rope diameter under adherence to the D/d=40 in accordance to EN81-1 [1] or by reduction of the D/d-ratio. Both measures have a big influence to rope lifetime and often underestimated. It is a well-known fact in traditional rope research that for steel wire ropes running over sheaves, the number of bending cycles decisively increases in proportion to an increase in the diameter ratio 'sheave to rope' (D/d) owing to decreasing bending stresses. Given an exemplary rope Filler 8x19 with a fibre core with a nominal rope diameter of d=12mm, the number of bending cycles is amplified by a factor 8, approximately, and this in proportion to an increase in the diameter ratio from D/d=25 to D/d=40 with a given safety factor of $\nu = 12$.

But: the market is asking for smaller sheave diameter, smaller rope diameter and lower motor moments. Additionally the lift manufacturers are asking for more freedom in construction issues eg. with the wish reducing the number of ropes which decreases the safety factor. This is totally out of the calculation methods of the standard. New challenges for rope lifetime are resulting from the more or less open D/d-ratio limits possible by certificates concerning the examination of conformity by notified bodies. This certificates are basing on series of endurance tests, statistical calculations with rrough statistical limitations. Important for the dialogue between rope manufacturer, lift manufacturer and user are the impression about the consequences by reducing or downsizing the dimensions of the rope drive. Here decision graphes as shown in [6] where the maximum number of trips are a function of the traction capacity and the safety factor. The number of trips are maximal possible but not guaranteed because other possible negative effects from transport, storage, handling, installation and maintenance have big influences to the rope performance.

5. Applications with reverse bendings

The steel wire ropes used in traction lifts are exposed to bending fatigue as they pass over the traction and deflection sheaves. As regards the bending sequence, attention must be paid to the occurrence of simple and reverse bending, which exert a greater or less impact on rope lifetime. In traction lifts reverse bendings occur in the steel wire ropes when running over the traction and deflection sheave where the traction sheave is arranged above or below next to the travel path. Reverse bending processes also have to be assumed in situations where an additional deflection sheave is needed to increase the wrapping angle to increase the traction capacity. Another reason is the change from a "big machine" to a machine with small traction sheave diameter. An additional deflection sheave with reverse bending and a horizontal rope length between the deflection sheaves which is in some cases critical for vibration effects. We call this the result of retrofitting with smaller dimensions on the machine side. Also the esthetical aspect of building without a machine room is another reason for this reverse bending because the space in the bottom of the building beside the shaft is used. In [8] a method for the calculation of the rope lifetime – better number of trips until discard – is described. The comparison between field observation and the calculation with the model showed high accuracy

6. Discard

The Discard is an essential safety-relevant requirement for safe operation. Visually detectable wire breaks on reference lengths and diameter reduction are of continual importance for steel wire ropes. Service time may well be a discard criterion in aforementioned far reaching tests. This fits only in special cases and applications.

Very often the lift manufacturers use plastic deflection sheaves instead of the steel or cast iron deflection sheaves. What are the effects? Using plastic deflection sheaves leads to an transmission of the highest stress in the contact zone rope - groove surface to the inner parts of the rope. The lifetime increases but the discard detection by wire breaks on the surface of the rope is not given in all cases. The increase in lifetime is lower than expected because the number of bends is normally higher than in other application eg in cranes where plastic sheaves show bigger advantages. In lifts the cost factor is dominant using plastic sheaves. The disadvantage discard criteria with plastic sheaves by wire break is a leck and often discussed. Normally the plastic sheaves are combined with a traction sheave of steel or cast iron. In [7] results showed that with this combination the discard determination is reliable.

7. "Soft facts" – further influences on rope lifetime

Until now we highlighted the different influences from the ropes and more and from the lift application parameters on lifetime of the ropes. The life of a running lift ropes starts earlier then in use only. Also very often underestimated are the influences of transport, storage, installation and maintenance. It´s very important keeping the ropes away from heat, water and dust during storage until installation. The ropes and in the case of master reels the handling must happen without damages. Here damages are also local deformation on several wires with the result in operation of a high number of wire breaks on short reference lengths. Installation must happen without eg untwisting the ropes to avoid damages of the rope and strand structure leading to unequal load carrying from wire to wire. During operation the ropes need a sufficient but also limited relubrication. Dry ropes have a low lifetime and produce very often rust as an irreversible fact.

8. Conclusion section

This paper examined the lifetime of steel wire ropes running over sheaves, the influences specific to lifts on rope lifetime – in particular the reduction of the D/d-ratio and the rope safety factor – and the method used to calculate rope lifetime in the application and with the relevant parameters. An examination against the background of the international standardization in lift engineering was given. Also an overview is given about the influences on rope lifetime and the reasons for that. Combined with the experiences with reverse bending operation and lifetime reduction by grooves found in a lot of field near tests here a new collection of several lifetime influences of steel ropes in traction lifts are bundled.

References

[1] DIN EN 81-1, Ausgabe:2000-05 Sicherheitsregeln für die Konstruktion und den Einbau von Aufzügen - Teil 1: Elektrisch betriebene Personen- und Lastenaufzüge

[2] Feyrer K 2015 Wire Ropes. Tension, Endurance, Reliability *SpringerVerlag 2015*

[3] Schiffner G 1999 Zur Ermittlung des Sicherheitsfaktors von Tragseilen. *Paper at Heilbronner Aufzugtage 1999*

[4] DIN EN 81-20/50 Sicherheitsregeln für die Konstruktion und den Einbau von Aufzügen – Prüfungen-Teil50: Konstruktionsregeln, Berechnungen und Prüfungen von Aufzugs-komponenten; German version EN81-50:2014

[5] Berner O 2011 Lebensdauer von Stahlseilen beim kombinierten Lauf über Treib- und Ablenkscheibenunterschiedlicher Rillenformen. *Dissertation Universität Stuttgart 2011*

[6] KP067-2 - certificates concerning the examination of conformity rope with and without reduction of number of trips DRAKO 250T 6,6.5 and 8mm, TÜV Sued 2013

[7] Vogel W 2010 Small Ropes – Aspects for Usage in Elevator Application. *Elevcon 2010 Lucerne/Switzerland*

[8] Vogel W 2010 The impact of reverse bending in the rope drives of traction lifts. *LiftReport 36(2010)2 pg. 14-16*

Dynamic behavior of elevator compensating sheave during buffer strike

S Watanabe[1], **T Okawa**[2]

[1] Advanced Technology R&D Center, Mitsubishi Electric Corp., 8-1-1 Tsukaguchi Honmachi, Amagasaki, Hyogo, Japan

[2] Inazawa Works, Mitsubishi Electric Corp., 1 Hishi-machi, Inazawa, Aichi, Japan

E-mail: `Watanabe.Seiji@ay.MitsubishiElectric.co.jp`

Abstract. This paper shows an elevator dynamic model that calculates the compensating sheave motion during a buffer strike. Our equivalent 2-degree-of-freedom vibration model of an elevator system, which consists of a car, a compensating sheave, and compensating ropes, can evaluate the dynamic tension of the compensating ropes caused by a buffer strike. The constraint force, which restricts the upward motion of the compensating sheave, is estimated from the dynamic rope tension. The constraint force is represented by the summation of two vibration modes and is the function of the limited distance of the compensating sheave's upward movement. Our formula, which evaluates the maximum constraint force, shows that a shorter limited distance of the compensating sheave increases the constraint force.

1. Introduction

Most elevators used in skyscrapers include compensating ropes to satisfy the balanced rope tension between the car and the counter-weight. The compensating ropes receive tension from a compensating sheave, which is installed at the bottom of the elevator shaft. Since the compensating sheave is only suspended by the compensating ropes, the sheave can move vertically while the car is traveling.

When the car can't stop at the highest floor during abnormal situations, the counter-weight can reduce its downward speed in 1g deceleration by a buffer strike for safe stops. During buffer strikes of the counter-weight, suspension ropes lose tension, and the car continues to ascend against gravity's force. The car's upward motion also induces the same upward motion against the compensating sheave. However, the sheave can't move beyond a limited distance for safety reasons, and its restricted upward motion induces higher tension to the compensating ropes.

This paper shows an elevator dynamic model that calculates the compensating sheave motion during a buffer strike. First, the transient behavior of the compensating ropes' tension and the compensating sheave's motion during a buffer strike is validated by the multi-body dynamic model of our elevator system. Its simulation model includes the effects of the buffer's reaction force, tension loss, and restrictions on the sheave's upward motion. Second, we used an equivalent 2-degree-of-freedom (2-DOF) vibration model to calculate the rope tension and the sheave motion. Finally, we explain the relation between the upward limit of the sheave and the rope tension by an equivalent 2-DOF model.

2. Multi-body dynamic model

Figure 1 shows the multi-body dynamic model of our elevator system. Since the suspension and compensating ropes are treated as continuous bodies, they are modeled by alternating the intensive mass and spring [1].

When the counter-weight strikes the buffer at the bottom of the elevator shaft, the car jumps and the compensating sheave also moves upward (figure 2). The compensating sheave's upward motion is restricted by a mechanical stopper. For safety reasons, the compensating sheave only rises to the limit of the mechanical stopper.

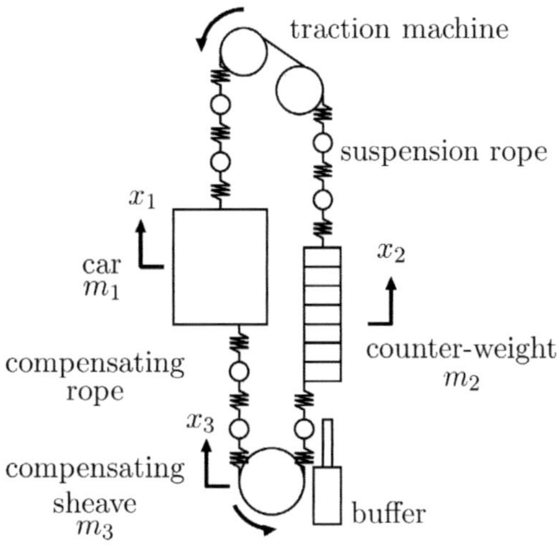

Figure 1. Elevator configuration.

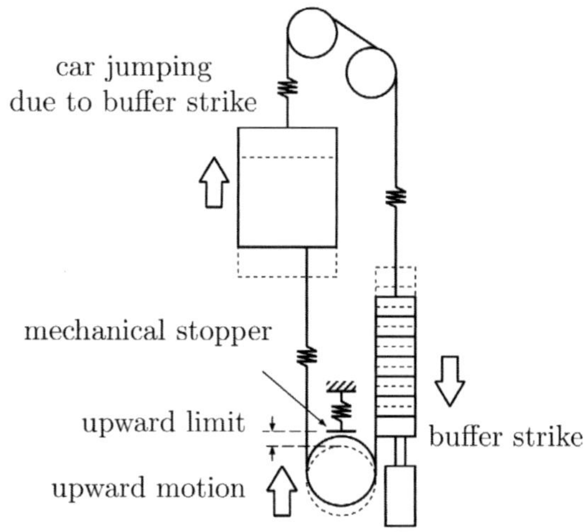

Figure 2. Sheave's upward motion.

As the upward motion is inhibited by the stopper, the compensating ropes experience high tension due to the car's continuous upward movement. The multi-body dynamic model in figure 1 is represented by the following equation of motion:

$$M\ddot{\boldsymbol{x}} + C\dot{\boldsymbol{x}} + K\boldsymbol{x} = \boldsymbol{F}, \tag{1}$$

where $\boldsymbol{x}$ is the vector of state variables. The vector contains the translational motion of car, counter-weight, compensating sheave and divided rope masses. It also contains the rotational motion of traction machine and compensating sheave.

When a rope tension imbalance on the sheave reaches the traction limit, the rope starts slipping. Otherwise, the rope's rotational speed on the sheave is constrained by the sheave speed due to the traction force.

In the case of buffer strike of the counter-weight, the counter-weight motion is constrained by the buffer motion.

These constraint motions of rope slip and buffer strike are evaluated by the following extended equation of motion:

$$\begin{bmatrix} M & E \\ E^T & 0 \end{bmatrix} \begin{bmatrix} \ddot{\boldsymbol{x}} \\ \Lambda \end{bmatrix} = \begin{bmatrix} \boldsymbol{F} - C\dot{\boldsymbol{x}} - K\boldsymbol{x} \\ f_e \end{bmatrix}, \tag{2}$$

where E and Λ are the constraint matrix and the constraint force vector as Lagrangian multipliers, respectively [2].

3. Equivalent model

3.1. Car and counter-weight motion during sheave's upward motion

The multi-body dynamic model in figure 1 can numerically calculate the transient rope tension during a buffer strike. However, since it remains unclear how the compensating sheave's motion affects the rope tension, we focus on the transient motion of the compensating ropes and the compensating sheave to evaluate it. To simplify the model in figure 1, we assume the following conditions:

- Before the counter-weight strikes the buffer, the car keeps rising at a constant speed and the counter-weight falls at the same constant speed.
- The suspension ropes lose tension when the car jumps due to the counter-weight striking the buffer and continues to climb against gravity.

Time difference t_0 from the counter-weight that strikes the buffer to the suspension ropes that lose tension is given by

$$t_0 = \frac{L_m}{v_m} \quad , \qquad v_m = \sqrt{\frac{E_m}{\rho_m}}, \tag{3}$$

where L_m is the suspension rope length of the counter-weight, E_m and ρ_m are Young's modulus and the density of the suspension rope. v_m is the propagation velocity in the rope axial direction.

The car speed is expressed by

$$\ddot{x}_1(t) = \begin{cases} 0 & (t < t_0) \\ -g & (t \geq t_0) \end{cases} \quad \longrightarrow \quad \dot{x}_1(t) = \begin{cases} V_0 & (t < t_0) \\ V_0 - g(t - t_0) & (t \geq t_0) \end{cases} , \tag{4}$$

where V_0 is the initial speed at time t which equals zero.

The counter-weight's speed is reduced by the buffer's resistance. Its average deceleration is 1g, but the typical deceleration pattern due to a buffer strike is approximated by the following linear function of time t. The counter-weight speed is given by the integration of the acceleration:

$$\ddot{x}_2(t) = \begin{cases} C_0(1 - t/t_2) & (t < t_2) \\ 0 & (t \geq t_2) \end{cases} \longrightarrow \dot{x}_2(t) = \begin{cases} -V_0 + C_0\left(t - \dfrac{t^2}{2t_2}\right) & (t < t_2) \\ 0 & (t \geq t_2) \end{cases} , \tag{5}$$

where t_2 is the stop time of the counter-weight and C_0 is the initial deceleration caused by the buffer strike.

3.2. Equation of motion related to car and compensating sheave

An equivalent 2-DOF vibration model of the compensating ropes is shown in figure 3. We assume that the car is in place on the top floor and the counter-weight is in place on the bottom floor. Table 1 defines the parameters.

The kinetic energies of car and compensating sheave are given by

$$T_1 = \frac{1}{2} m_1 \dot{x}_1^2 \quad , \quad T_3 = \frac{1}{2} m_3 \dot{x}_3^2. \tag{6}$$

The kinetic energy of the distributed rope mass is given on the assumption of the rope's linear displacement. Therefore, the kinetic energy in any short part of the rope is given by

$$\Delta T_r = \frac{1}{2} \rho dx \left\{ \frac{x}{L} (\dot{x}_1 - \dot{z}) + \dot{z} \right\}^2 . \tag{7}$$

The total kinetic energy of the rope is given by integrating the above equation:

$$T_r = \int_0^L \Delta T_r = \frac{1}{2} \frac{m_r}{3} \left\{ \dot{x}_1^2 + \dot{x}_1 \dot{z} + \dot{z}^2 \right\} \quad , \qquad m_r = \rho L. \tag{8}$$

Table 1. Model parameters.

x_1	car displacement
x_2	counter-weight displacement
x_3	compensating sheave displacement
z	bottom rope displacement
m_1	car weight
m_3	compensating sheave weight
k	stiffness of compensating rope
ρ	line density of compensating rope
L	compensating rope length

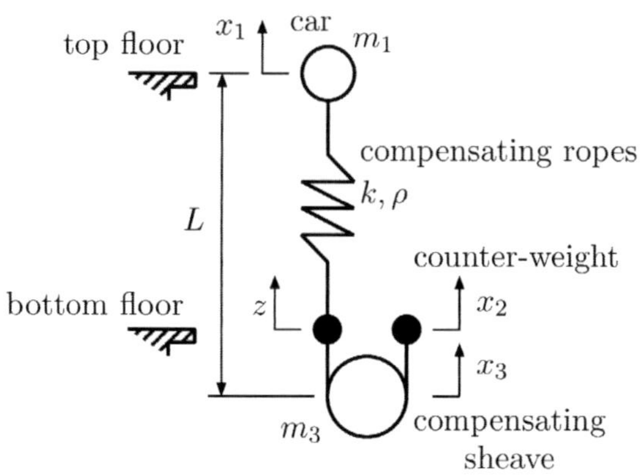

Figure 3. Equivalent 2-DOF model.

The potential energy of compensating rope is given by

$$U = \frac{1}{2}k(x_1 - z)^2. \tag{9}$$

Variable z is the function of x_2 and x_3:

$$z = 2x_3 - x_2. \tag{10}$$

The equations of motion related to the car and the compensating sheave are given by Lagrange formulation of the total kinetic energy and the potential energy of Eqs. (6), (8) and (9):

$$x_1: \quad \left(m_1 + \frac{m_r}{3}\right)\ddot{x}_1 + \frac{m_r}{3}\ddot{x}_3 + k(x_1 + x_2 - 2x_3) = \frac{m_r}{6}\ddot{x}_2 - \left(m_1 + \frac{1}{2}m_r\right)g \tag{11}$$

$$x_3: \quad \frac{m_r}{3}\ddot{x}_1 + \left(m_3 + \frac{4m_r}{3}\right)\ddot{x}_3 - 2k(x_1 + x_2 - 2x_3) = \frac{2m_r}{3}\ddot{x}_2, \tag{12}$$

where the suspension ropes lose tension after the counter-weight strikes the buffer. Eqs. (11) and (12) have coupling effects against the inertia and stiffness terms.

3.3. Constraint force of compensating sheave

The following equation highlights the relationship between x_2 and x_3 by deleting the stiffness term from Eqs. (11) and (12) with substituting $-g$ for $\ddot{x}_1$, related to Eq. (4):

$$\ddot{x}_3 = \frac{m_r}{m_3 + 2m_r}\ddot{x}_2 = \beta\ddot{x}_2. \tag{13}$$

If total rope weight m_r is much heavier than compensating sheave m_3, β is close to 0.5. Therefore the compensating sheave's acceleration $\ddot{x}_3$ is half of counter-weight $\ddot{x}_2$.

From Eq. (13), we can estimate time t_3 at which the compensating sheave reaches the limit of the upward motion.

We introduce a new variable to evaluate the rope elongation: $y = x_1 + x_2 - 2x_3$. After time t_3 of the upward limit of the compensating sheave, the equation of motion related to the car in Eq. (11) is modified, because the compensating sheave's acceleration $\ddot{x}_3$ equals zero:

$$\left(m_1 + \frac{m_r}{3}\right)\ddot{y} + ky = \left(m_1 + \frac{m_r}{2}\right)(\ddot{x}_2 - g). \tag{14}$$

Rope elongation $y(t)$ is given by

$$y(t) = C_1 \cos\omega(t - t_3) + C_2 \sin\omega(t - t_3) + \gamma(\ddot{x}_2 - g) \tag{15}$$

$$\omega = \sqrt{\frac{k}{m_1 + m_r/3}}\ , \ \gamma = \frac{1}{k}\left(m_1 + \frac{m_r}{2}\right)\ , \ C_1 = -\gamma(\ddot{x}_2(t_3) - g)\ , \ C_2 = \frac{1}{\omega}\{-\gamma\dddot{x}_2(t_3) + \dot{y}(t_3)\},$$

where rope elongation speed $\dot{y}$ at time t_3 is given by

$$\dot{y}(t_3) = \dot{x}_1(t_3) + \dot{x}_2(t_3) - 2\dot{x}_3(t_3) = \dot{x}_1(t_3) + \dot{x}_2(t_3) \qquad (\because \dot{x}_3(t_3) = 0). \tag{16}$$

$\dot{x}_1(t_3)$ and $\dot{x}_2(t_3)$ are calculated by Eqs. (4) and (5).

The equation of motion related to the compensating sheave in Eq. (12) is modified using y:

$$\frac{m_r}{3}\ddot{y} + (m_3 + 2m_r)\ddot{x}_3 - m_r\ddot{x}_2 - 2ky = 0. \tag{17}$$

When the compensating sheave reaches the limit of the upward motion, $\ddot{x}_3$ is zero at time t_3 and the compensating sheave's upward motion is restricted. Therefore the equation of motion in Eq. (17) includes constraint force λ to satisfy the constraint equation:

$$\begin{cases} \frac{m_r}{3}\ddot{y} + (m_3 + 2m_r)\ddot{x}_3 - m_r\ddot{x}_2 - 2ky = -\lambda \\ \ddot{x}_3 = 0 \end{cases} \longrightarrow \quad -\lambda = \frac{m_r}{3}\ddot{y} - m_r\ddot{x}_2 - 2ky. \tag{18}$$

Substituting Eqs. (14) and (15) for Eq. (18),

$$\lambda(t) = -(2m_1 + m_r)\left\{-\omega^2\left(C_1\cos\omega(t - t_3) + C_2\sin\omega(t - t_3)\right) + g\right\} + 2(m_1 + m_r)\ddot{x}_2(t). \tag{19}$$

Constraint force λ in Eq. (19) is twice the compensating ropes' tension, which includes the rope's inertia force.

When the mechanical stopper inhibits the compensating sheave from rising, the compensating ropes receive an impact force that axially propagates the rope vibration:

$$F(t) = F_s\{1 - \cos\omega_r(t - t_3)\} \quad , \qquad F_s = (m_3 + 2m_r)\ddot{x}_3(t_3), \tag{20}$$

where impact force F_s corresponds to the second term in Eq. (17) and ω_r is the eigen frequency of the rope axial vibration:

$$\omega_r = \frac{\pi}{L}\sqrt{\frac{EA}{\rho}}, \tag{21}$$

where E and A are Young's modulus and the cross section area of the compensating rope, respectively.

Finally, maximum force F_{max}, which restricts the compensation sheave's upward movement, is given by the summation of Eqs. (19) and (20). As the frequency of impact force $F(t)$ is much higher than the frequency of $\lambda(t)$, the maximum force F_{max} occurs at the peak time t_f of $F(t)$:

$$C_f(t) = \lambda(t) + F(t) \quad \longrightarrow \quad F_{max} = C_f(t_f) = \lambda(t_f) + F(t_f), \tag{22}$$

where $t_f = t_3 + \dfrac{(2n-1)\pi}{\omega_r}$, $n = 1, 2, \cdots$.

4. Simulation results

Next we compare the formulas in Section 3 with the numerical simulation by the multi-body dynamic model in Section 2. The simulation model in Eq. (2) consists of 46 degrees of freedom and it is solved by Runge-Kutta method. Figure 4 shows the time response of the acceleration against the car and the counter-weight. The thin lines correspond to Eqs. (4) and (5).

The upward motion of the compensating sheave is calculated by Eq. (13) (figure 5). As the displacement is normalized by the upward limit, the vertical axis in figure 5 is dimensionless.

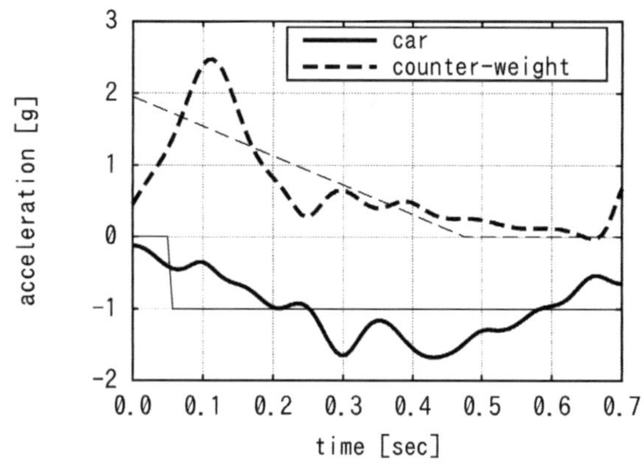

Figure 4. Time response of acceleration.

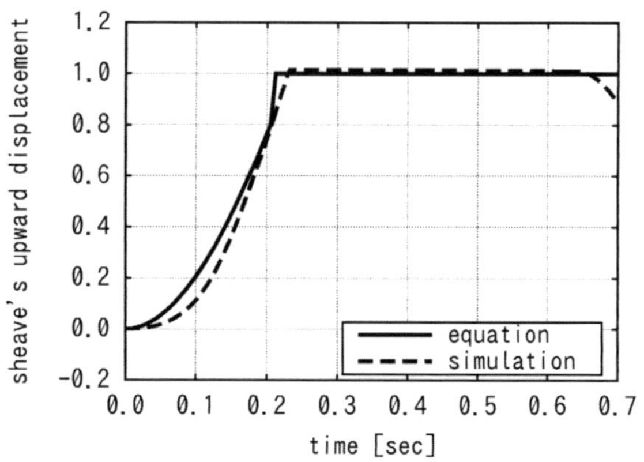

Figure 5. Normalized sheave's upward motion.

The time response of the constraint force in Eqs. (19) and (22) is shown in figure 6. The constraint force is normalized by the maximum force of the simulation result, therefore the vertical axis is represented as a dimensionless value. The solid line corresponds to Eq. (22) and the thin dotted line represents Eq. (19).

As shown in figure 5, time t_3 for reaching the upward limit is close to the simulation result. The transient constraint force in Eq. (22) also resembles the simulation result in figure 6. Therefore we conclude that the formula in Eq. (22) is appropriate to evaluate the maximum constraint force.

To investigate the effect of the upward limit against the constraint force, the upward limit is shifted $\pm$50mm in Eq. (22). Figure 7 shows that a shorter upward limit causes a larger constraint force.

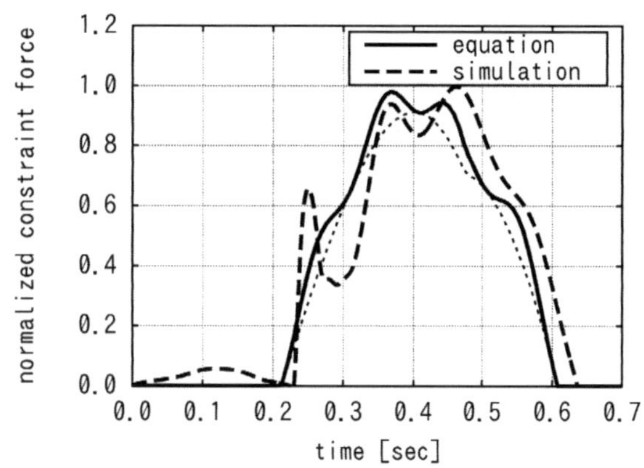

Figure 6. Constraint force.

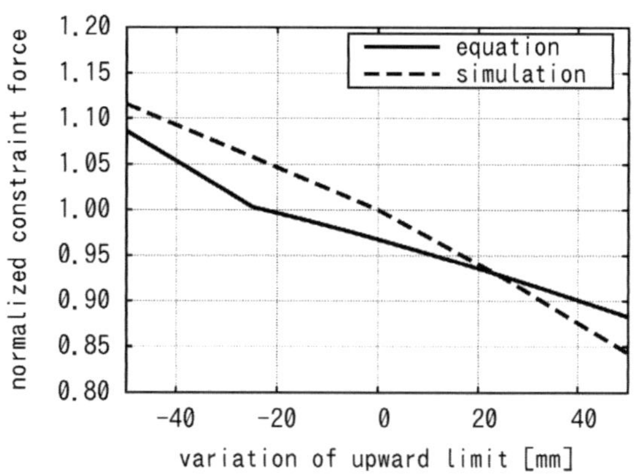

Figure 7. Effect of upward limit.

5. Conclusions

We evaluated the constraint force against a compensating sheave with an equivalent 2-DOF vibration model and validated our formula based on the 2-DOF model by numerical simulation. Our formula suggests that a shorter upward limit causes a larger constraint force that is represented by the summation of two vibration modes.

References

[1] Watanabe S, Okawa T, Nakazawa D and Fukui D 2013 *Proc. of MoSS 2013*
[2] Watanabe S, Yumura T, Hayashi Y and Takigawa Y 2002 *Proc. of ACMD'02*

Journal of Physics: Conference Series **721** (2016) 012016 doi:10.1088/1742-6596/721/1/012016

Vibrations of beams and rods carrying a moving mass

X W Zhao[1,2], G H M van der Heijden[1], Z D Hu[2]

[1] Department of Civil, Environmental & Geomatic Engineering, University College London, London, UK,

[2] School of Aerospace Engineering and Applied Mechanics, Tongji University, Shanghai, People's Republic of China

E-mail: xingwei.zhao@ucl.ac.uk, g.heijden@ucl.ac.uk, zdhu@tongji.edu.cn

Abstract. We study the vibration of slender one-dimensional elastic structures (beams, cables, wires, rods) under the effect of a moving mass or load. We first consider the classical small-deflection (Euler-Bernoulli) beam case, where we look at tip vibrations of a cantilever as a model for a barreled launch system. Then we develop a theory for large deformations based on Cosserat rod theory. We illustrate the effect of moving loads on large-deformation structures with a few cable and arch problems. Large deformations are found to have a resonance detuning effect on the cable. For the arch we find different failure modes depending on its depth: a shallow arch fails by in-plane collapse, while a deep arch fails by sideways flopping. In both cases the speed of the traversing load is found to have a stabilising effect on the structure, with failure suppressed entirely at sufficiently high speed.

1. Introduction

The problem of a continuously distributed system carrying a moving concentrated mass has broad applications in mechanics and engineering, including space tethers, satellite antennas, launch systems, robotic arms [1], cranes, flexible manipulators [2], high-speed train railroads and highway bridges with moving vehicles [3]. For an overview of the sizable early literature on the vibration theory of moving mass and load problems we refer to Fryba's monograph [4].

The classical example of a moving mass problem is the idealisation of a vehicle-bridge system. In this case the moving vehicle is usually treated as a moving force, or load, of constant magnitude, while the bridge is modelled as a simply-supported beam. This problem is therefore more accurately described as a moving load problem [3]. The moving load assumption does not take into account the inertial forces of the moving mass and of the interactions between the moving mass and the continuous beam. The moving load treatment is therefore insufficient when the gravitational and inertial effects of the moving mass are not negligible compared to the mass of the structure. In [5] it was shown that the moving load solution is not an upper bound for the moving mass solution for the (simply-supported and clamped-clamped) boundary conditions considered there.

Ting et al. [6] were among the first to consider the inertial forces, including centrifugal and Coriolis effects, of the moving mass as an interaction between the mass and the beam. Further studies of the moving mass problem include [7, 8, 9]. All these works restrict themselves to beams undergoing small deformations (in [8] moderately large deflections are considered) and all consider only two-dimensional bending vibrations. With the current drive to use thinner and lighter materials, in order to save material and reduce costs, large deformations become increasingly important.

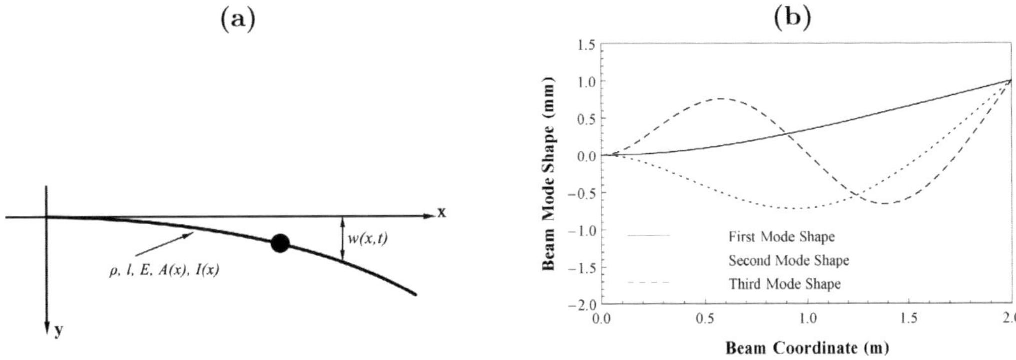

Figure 1. (a) Cantilever beam carrying a moving mass. (b) The first three mode shapes of the cantilever beam.

This paper first briefly reviews the small-deformation theory and then presents early results of our programme to develop a theory for slender elastic rods undergoing arbitrary deformations (under small strains) while carrying a moving load or mass. The small-deformation beam case is motivated by launch systems and studies the effects of a moving mass on the tip vibration of a cantilever beam, important for the projectile motion of the released mass. We have previously considered tapered beams [10], but here consider beams with uniform (circular) cross-section. The large-deformation rod case is motivated by cable-mass systems. We show some preliminary results in which a load (weight) traverses a cable or arch at constant speed, as is customary in the classical small-deformation case. Such prescribed motion of the mass will be unrealistic in certain application and a full large-deformation theory will ultimately have to describe the two-way coupling between rod and mass: the mass affects the deformation of the rod but the deformed rod will also affect the speed of the mass as it is forced to track the shape of the rod (under gravity). This is work in progress.

2. Beam with a moving mass

The beam is assumed to undergo small deflections. We can then write $x(t)$ for the position of the mass along the axis of the beam and $w(x(t),t)$ for the position of the mass in space (see Fig. 1(a)). The vertical velocity and acceleration of the moving mass are then computed as

$$\frac{\mathrm{d}w}{\mathrm{d}t} = v(t)\frac{\partial w}{\partial x} + \frac{\partial w}{\partial t}, \qquad \frac{\mathrm{d}^2 w}{\mathrm{d}t^2} = v^2(t)\frac{\partial^2 w}{\partial x^2} + 2v(t)\frac{\partial^2 w}{\partial x \partial t} + a(t)\frac{\partial w}{\partial x} + \frac{\partial^2 w}{\partial t^2}. \tag{1}$$

The first and second terms of the acceleration are the centrifugal and Coriolis terms, respectively.

We model the beam as an Euler-Bernoulli beam. The action functional for use in Hamilton's principle is

$$S = \int_{t_1}^{t_2} \left\{ \int_0^L \left[\frac{1}{2}\rho A(x)\left(\frac{\partial w}{\partial t}\right)^2 - \frac{1}{2}EI(x)\left(\frac{\partial^2 w}{\partial x^2}\right)^2 + \rho A(x) g w \right] \mathrm{d}x + \frac{1}{2}M\left(\frac{\mathrm{d}w}{\mathrm{d}t}\right)^2 + Mgw \right\} \mathrm{d}t. \tag{2}$$

Here L is the length, ρ the density, A the cross-sectional area, I the second moment of area about the z-axis and E Young's modulus. The mass M slides along the length of the beam with local speed $v(t)$ and acceleration $a(t)$.

We use a Galerkin approximation to spatially discretise the problem and write the beam deflection as a truncated series $w(x,t) = \sum_{i=1}^{n} \Phi_i(x)Y_i(t)$, where the $\Phi_i(x)$ are (the first n) mode shapes of the free beam and the $Y_i(t)$ are time-dependent dimensionless coefficients. After

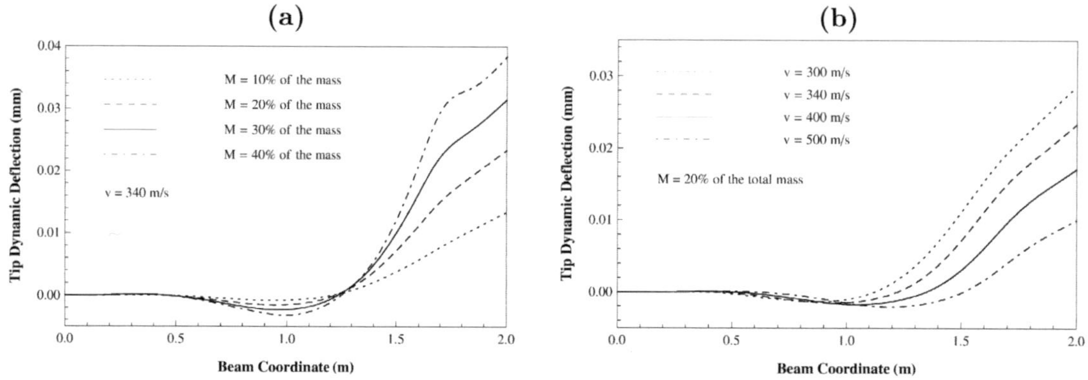

Figure 2. (a) Tip deflections at various values of M for different travelling velocities v. The 'beam coordinate' is the instantaneous position along the beam of the travelling mass. (b) Tip deflections at various values of v for different travelling masses M.

substitution of this series expansion into Eq. (2) the integrand of S becomes a function of Y_i and $\dot{Y}_i$ and the application of Hamilton's principle gives the Euler-Lagrange equations in the form

$$\mathbf{M}(t)\ddot{\mathbf{Y}}(t) + \mathbf{C}(t)\dot{\mathbf{Y}}(t) + \mathbf{K}(t)\mathbf{Y}(t) = \mathbf{P}(t), \tag{3}$$

where $\mathbf{Y}(t) = [Y_1(t), Y_2(t), \dots, Y_n(t)]^T$ and $\mathbf{M}$, $\mathbf{K}$ and $\mathbf{C}$ are the mass, stiffness and 'damping' matrices, respectively (note that $\mathbf{C}$ comes from the Coriolis effect of the moving mass and is antisymmetric and therefore does not actually represent damping in the sense that there is no dissipation). $\mathbf{M}$ is symmetric but $\mathbf{K}$ varies with the beam coordinate $x(t)$ as the mass moves along the beam and as a result is nonsymmetric. $\mathbf{P}$ is the load vector (due to gravity).

Fig. 1(b) shows the first three mode shapes of the cantilever beam. Here we have used the normalisation that the absolute tip deflections $|\Phi_i(L)|$ are 1.

Example. We consider a hollow steel beam ($E = 2.0 \times 10^{11}\,\mathrm{Pa}, \rho = 7850\,\mathrm{kg/m^3}$) of length $L = 2\,\mathrm{m}$, outer radius $R = 0.1\,\mathrm{m}$ and inner radius $r = 0.05\,\mathrm{m}$. The magnitude (M) and the (uniform) speed (v) of the moving mass will be varied. We focus on the effect of these parameters on beam tip and mass deflections as the beam is excited by the travelling mass. We take initial conditions where the mass starts moving when the beam is in the static equilibrium configuration and compute expansions of the beam deflection in terms of the first three mode shapes Φ_i, i.e., we take $n = 3$.

Solutions, obtained using the fourth-order Runge-Kutta method, are given in the dynamic tip deflection plots of Fig. 2 for four different velocities v and for four different values of M, specified as the percentage of the total mass of the beam-mass system. In each plot the tip deflection $w(L,t)$ of the beam is given as a function of the instantaneous position $x(t)$ of the moving mass. We call this the beam coordinate. Since the speed is constant it can be interpreted as a time coordinate. When $x(t) = L = 2$, the mass has reached the end of the beam.

We find that for the parameters used the tip vibrations are very small. At relatively high speeds the mass movement triggers the beam tip to vibrate. Generally, the higher the speed v and the larger the mass M, the higher the vibration frequency of the beam tip. The amplitude of delfection reduces as the speed v increases.

3. Large deformations – Cosserat rod formulation

Let $\boldsymbol{r}(s,t) : [0, L] \times \mathbb{R} \to \mathbb{R}^3$ be a smooth space curve of length L describing the centerline of the rod. Further, let $\{\boldsymbol{d}_1(s,t), \boldsymbol{d}_2(s,t), \boldsymbol{d}_3(s,t)\}$ be a frame of orthonormal directors such that $\boldsymbol{d}_1$

and $\boldsymbol{d}_2$ span the cross-section of the rod, pointing in its principal directions, and $\boldsymbol{d}_3 = \boldsymbol{d}_1 \times \boldsymbol{d}_2$ is perpendicular to it. Orthonormality of the frame implies the existence of two axial vectors $\boldsymbol{\kappa}$ (the curvature vector) and $\boldsymbol{\omega}$ (the spin vector) such that

$$\partial_s \boldsymbol{d}_k = \boldsymbol{\kappa} \times \boldsymbol{d}_k, \qquad \partial_t \boldsymbol{d}_k = \boldsymbol{\omega} \times \boldsymbol{d}_k \qquad (k = 1, 2, 3). \tag{4}$$

The force and moment equilibrium equations are given by

$$\partial_s \boldsymbol{n} + \boldsymbol{f} = \rho A \partial_{tt} \boldsymbol{r}, \qquad \partial_s \boldsymbol{m} + \partial_s \boldsymbol{r} \times \boldsymbol{n} + \boldsymbol{l} = \rho \partial_t (\boldsymbol{I}\boldsymbol{\omega}), \tag{5}$$

where $\boldsymbol{n}$ and $\boldsymbol{m}$ are the resultant contact force and contact couple acting at the centreline at s, and $\boldsymbol{f}$ and $\boldsymbol{l}$ are external forces and moments acting on the rod [11]. $\boldsymbol{I}$ is the moment of inertia tensor, A is the area of the cross-section and ρ is the density.

We take the rod to be inextensible and unshearable. Then the director $\boldsymbol{d}_3$ coincides with the tangent to the centreline of the rod, i.e., $\partial_s \boldsymbol{r} = \boldsymbol{d}_3$. If we also introduce the centreline velocity $\boldsymbol{u} = \partial_t \boldsymbol{r}$, then the identity $\partial_t \partial_s \boldsymbol{r}(s,t) = \partial_s \partial_t \boldsymbol{r}(s,t)$ can be written as

$$\partial_t \boldsymbol{d}_3 = \partial_s \boldsymbol{u}. \tag{6}$$

Similarly, the identity $\partial_t \partial_s \boldsymbol{d}_k(s,t) = \partial_s \partial_t \boldsymbol{d}_k(s,t)$, on using Eq. (4), leads to the curvature-spin compatibility equation

$$\partial_s \boldsymbol{\omega} = \partial_t \boldsymbol{\kappa} + \boldsymbol{\kappa} \times \boldsymbol{\omega}. \tag{7}$$

When expressed in the director frame, Eqs (5), (6), (7) give us the following equations (using a sans-serif font for triples of components and denoting $\boldsymbol{I}$ on principal axes by $\mathsf{I} = \mathrm{diag}(I_1, I_2, J)$):

$$\begin{aligned} \rho A \partial_t \mathsf{u} &= \partial_s \mathsf{n} + \kappa \times \mathsf{n} - \rho A(\omega \times \mathsf{u}) + \mathsf{f}, \\ \rho \mathsf{I} \partial_t \omega &= \partial_s \mathsf{m} + \kappa \times \mathsf{m} + \mathsf{d}_3 \times \mathsf{n} - \rho(\omega \times \mathsf{I}\omega) + \mathsf{l}, \\ \partial_t \kappa &= \partial_s \omega + \kappa \times \omega, \\ 0 &= \partial_s \mathsf{u} + \kappa \times \mathsf{u} - \omega \times \mathsf{d}_3. \end{aligned} \tag{8}$$

We finally introduce linear constitutive relations between the curvatures κ and the (body) moments m: $\mathsf{m} = \bar{\mathsf{K}}\kappa$, where $\bar{\mathsf{K}} = \mathrm{diag}(B_1, B_2, C)$, with $B_1 = EI_1$ and $B_2 = EI_2$ the principal bending stiffnesses and $C = GJ$ the torsional stiffness, E and G being Young's modulus and the shear modulus respectively. These relations can be used to eliminate m in Eq. (8) in favour of κ, thus giving us 12 equations for the body components $(\mathsf{u}, \omega, \kappa, \mathsf{n}) =: \boldsymbol{y}$ in the form

$$\boldsymbol{M} \partial_t \boldsymbol{y}(s,t) + \boldsymbol{K} \partial_s \boldsymbol{y}(s,t) + \boldsymbol{F}(s,t) = \boldsymbol{0}, \tag{9}$$

where $\boldsymbol{M}$ and $\boldsymbol{K}$ are mass and stiffness matrices (constant in the case of a uniform rod). All nonlinear terms in Eq. (8) are contained in the load vector $\boldsymbol{F}$. The same formulation is used in [12, 13].

To solve the equations numerically we apply the generalised α-method [14] for both spatial and temporal discretisation. This yields the discrete form of Eq. (9) as follows:

$$\begin{aligned} &\boldsymbol{M}\{(1-\alpha_t)[(1-\alpha_s)\partial_t \boldsymbol{y}_j^i + \alpha_s \partial_t \boldsymbol{y}_{j-1}^i] + \alpha_t[(1-\alpha_s)\partial_t \boldsymbol{y}_j^{i-1} + \alpha_s \partial_t \boldsymbol{y}_{j-1}^{i-1}]\} \\ &+\boldsymbol{K}\{(1-\beta_t)[(1-\beta_s)\partial_s \boldsymbol{y}_j^i + \beta_s \partial_s \boldsymbol{y}_{j-1}^i] + \beta_t[(1-\beta_s)\partial_s \boldsymbol{y}_j^{i-1} + \beta_s \partial_s \boldsymbol{y}_{j-1}^{i-1}]\} \\ &+\{(1-\beta_t)[(1-\beta_s)\boldsymbol{F}_j^i + \beta_s \boldsymbol{F}_{j-1}^i] + \beta_t[(1-\beta_s)\boldsymbol{F}_j^{i-1} + \beta_s \boldsymbol{F}_{j-1}^{i-1}]\} = \boldsymbol{0}. \end{aligned} \tag{10}$$

In this equation the superscript indicates the time step while the subscript indicates the spatial step, i.e., the node of the discretised curve. The s and t derivatives are approximated as

$$\partial_t \boldsymbol{y}^i = \frac{\boldsymbol{y}^i - \boldsymbol{y}^{i-1}}{\gamma_t \Delta t} - \frac{1-\gamma_t}{\gamma_t} \partial_t \boldsymbol{y}^{i-1}, \qquad \partial_s \boldsymbol{y}_j = \frac{\boldsymbol{y}_j - \boldsymbol{y}_{j-1}}{\gamma_s \Delta s} - \frac{1-\gamma_s}{\gamma_s} \partial_s \boldsymbol{y}_{j-1}, \tag{11}$$

where Δs and Δt are the spatial step and time step, respectively. The six numerical parameters α_s, β_s, γ_s and α_t, β_t, γ_t can be tuned to obtain desirable accuracy and stability properties. We choose the value 1/2 for all six parameters, which gives a stable second-order scheme without numerical damping. The final system of algebraic equations is solved with the global Newton code NLEQ1 [15].

To compute rod shapes in space we need to postprocess the solution $\boldsymbol{y}$ by subsequently solving either of the equations in (4) and integrating the centreline equation $\partial_s \boldsymbol{r} = \boldsymbol{d}_3$ or the velocity $\boldsymbol{u} = \partial_t \boldsymbol{r}$ to obtain $\boldsymbol{r}$. Let $\{\boldsymbol{i}, \boldsymbol{j}, \boldsymbol{k}\}$ be a fixed coordinate frame in space. We define R as the matrix whose columns are the frame vectors $\boldsymbol{i}$, $\boldsymbol{j}$, $\boldsymbol{k}$ represented in the body frame. R is thus the rotation matrix from the inertial to the body frame. Assuming we choose to use the second equation in (4) (i.e., time integration), the rotation matrix is updated according to

$$R(t + \Delta t) = e^{-\hat{\omega}\Delta t} R(t), \qquad (12)$$

with initial condition $R(0) = I$ if the body frame at $t = 0$ is aligned with the fixed frame [16]. Here, $\hat{\omega}$ is the usual skew-symmetric matrix corresponding to the axial vector ω. The exponential is conveniently computed by using the Rodrigues formula (valid for any skew-symmetric matrix $\hat{\mathsf{x}}$)

$$e^{-\hat{\mathsf{x}}} = I - \hat{\mathsf{n}} \sin|\mathsf{x}| + \hat{\mathsf{n}}^2(1 - \cos|\mathsf{x}|), \qquad \mathsf{n} = \frac{\mathsf{x}}{|\mathsf{x}|}. \qquad (13)$$

Having found R, we finally obtain the centreline $\boldsymbol{r}$ by integrating the velocity vector:

$$\partial_t \boldsymbol{r} = \boldsymbol{u} = R^{\mathrm{T}} \mathsf{u}. \qquad (14)$$

The integrations in these postprocessing steps are carried out using the trapezoidal rule to maintain second-order accuracy.

The rotation matrix R is also used to transform any external forces or moments specified in the fixed frame (for instance gravity) to the body frame for inclusion in Eq. (8):

$$\mathsf{f} = R\boldsymbol{f}, \qquad \mathsf{l} = R\boldsymbol{l}. \qquad (15)$$

Boundary conditions, if specified in the fixed frame, may similarly be transformed to the body frame.

4. Applications

We apply our formulation to a few structures that are initially planar but may suffer out-of-plane instabilities under the action of a moving point load F. We orient our fixed coordinate system $\{\boldsymbol{i}, \boldsymbol{j}, \boldsymbol{k}\}$ such that $\boldsymbol{j}$ is pointing vertically up, $\boldsymbol{k}$ is pointing horizontally along the end-to-end axis of the rod (in the direction of increasing s) and $\boldsymbol{i}$ is pointing out of the plane of the rod. We denote the components of the centreline of the rod, $\boldsymbol{r}$, in this fixed frame by (x, y, z) and choose the left end point to be at $(x, y, z) = (0, 0, 0)$. The external loads are gravity and the (vertical) moving load F, so we set $\mathsf{f} = -(\rho A g + F)\mathsf{j} = -(\rho A g + F)R\boldsymbol{j}$ and $\mathsf{l} = \mathsf{0}$ in Eq. (8), where g is the gravitational acceleration.

In all the following examples the rod is taken to be uniform, of length $L = 1\,\mathrm{m}$ and radius $r = 0.005\,\mathrm{m}$, while $E = 2.5 \times 10^8\,\mathrm{Pa}$, $\rho = 1500\,\mathrm{kg/m^3}$ and $g = 9.8\,\mathrm{m/s^2}$.

4.1. Hanging cable

Consider a cable hanging under gravity. Its end tangents are horizontal and prevented from rotation but the right end is free to move along the end-to-end axis (see true views in Figs 3(a) and (c)). At time $t = 0$ a vertical load of magnitude F starts to travel from the left to the right end with constant speed $v = \mathrm{d}\sigma/\mathrm{d}t$, where σ is the time-like 'rod coordinate', i.e., the

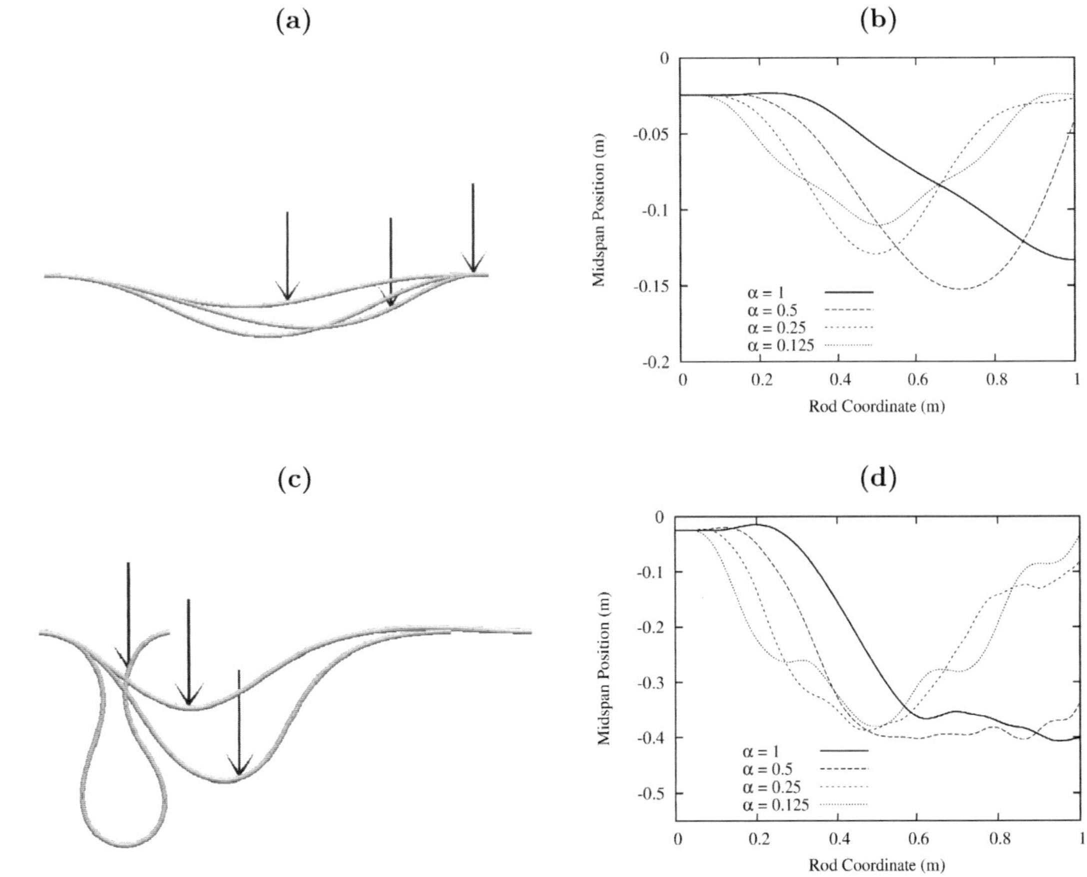

Figure 3. Cable shapes and vertical midspan positions for two different loads F moving at various speeds v given as a fraction α of the critical speed: (a,b) small load, $F = 2.0\,\mathrm{N}$; (c,d) large load, $F = 20.0\,\mathrm{N}$.

instantaneous arclength position of the travelling load. This, for instance, models the motion of inspection robots along bridge cables or high-voltage power lines [17]. We consider two cases: a small load ($F = 2.0\,\mathrm{N}$) and a large load ($F = 20.0\,\mathrm{N}$). Good numerical results are obtained by taking $\Delta s = 0.01$ and $\Delta t = \Delta s/v$, the latter having the effect that the cable solution is sampled at times when the load is applied exactly at a node.

Fig. 3 shows oscillations in the cable's vertical midspan position $y(L/2)$ plotted against the rod coordinate σ for various speeds v given as a fraction α of the first critical speed v_1. For small vibrations this critical speed is the speed required to cross the cable in half a period of its first bending mode, i.e., $v_1 = \omega_1 L/\pi$, where for fixed-fixed boundary conditions the first natural frequency ω_1 is given by $\omega_1 = (4.7300/L)^2\sqrt{EI/\rho A}$. We observe that in the case of a small load (Fig. 3(a,b)) there is a resonance phenomenon in the sense that the maximum amplitude of vibrations occurs for speeds near the resonance speed ($\alpha = 1$). The rod effectively behaves as a stiff beam and the results agree with those for an Euler-Bernoulli beam found in [18]. In the case of a large load (Fig. 3(c,d)) we observe a significant detuning effect: the rod behaves as a flexible cable with no resonance effect shown. The dimensionless parameter controlling the effective stiffness/flexibility of the loaded cable is EI/FL^2.

4.2. Hinged arch

Next we consider a constant vertical load traversing an arch. The initial arch is formed by buckling (upwards) an initially straight hinged-hinged rod under a quasistatically increasing compressive horizontal end load. The end-to-end distance is then fixed and gravity is 'switched on' by gradually increasing g from 0 to $9.8\,\mathrm{m/s^2}$ (the end force $\boldsymbol{n}(L)\cdot\boldsymbol{k}$ varies passively in this process). We do these initial parameter continuations with a separate statics code, so at this point we have a pure statics solution with all velocities and angular velocities equal to zero. We consider two types of arch, a shallow and a deep one. The shallow arch has its end-to-end distance fixed at $\Delta z = 0.864545\,\mathrm{m}$ and carries a load $F = 2.0\,\mathrm{N}$, while the deep arch has end-to-end distance $\Delta z = 0.529689\,\mathrm{m}$ and carries a load $F = 3.8\,\mathrm{N}$. The two arches have different failure modes under these loads when applied statically, as illustrated in Fig. 4: the shallow arch fails by in-plane collapse (described by the fold in the figure), while the deep arch fails by out-of-plane sideways flopping (described by a pitchfork bifurcation, which is followed by a restabilising reverse pitchfork bifurcation).

Instead of the static load we now consider dynamic loading, letting F move from the left end with constant speed v. Results are shown in Figs 5–7. In Fig. 5, for the shallow arch, we see that for small speed v the arch fails close to the point of quasistatic collapse, while at larger v collapse is significantly delayed and for speeds larger than $v = 2.6\,\mathrm{m/s}$ no collapse occurs at all. As a measure for this collapse we show in Fig. 5(a) the angular velocity $\boldsymbol{\omega}(L)\cdot\boldsymbol{i}$ about the hinge at $s = L$: its value first stays small as the load moves to the right and then suddenly goes sharply negative (corresponding to a rapid clockwise rotation of the right end tangent) as the arch snaps under the load. Shapes for $v = 0.5\,\mathrm{m/s}$ are shown in Fig. 5(b). Fig. 5(c) illustrates how the discontinuity in the internal force, as a result of the point load F, travels along the rod. Our numerical scheme has no problem with these discontinuities. Fig. 5(d) illustrates the stabilising effect of large v: for $v = 2.6\,\mathrm{m/s}$ the arch still significantly deforms under the travelling load, but no sudden jump occurs and the right end tangent does not change sign, while for $v = 20\,\mathrm{m/s}$ the arch deforms hardly at all under the moving load. The numerical parameters used in these shallow-arch runs are: $\Delta s = 0.00125$ for $v = 0.02$ and $\Delta s = 0.01$ for all other speeds, and again $\Delta t = \Delta s/v$.

Fig. 6, for the deep arch, shows that for sufficiently small speed v the arch goes unstable close to the point of quasistatic instability, with the arch oscillating about the stable out-of-plane statics solution (Fig. 6(a,b) for $v = 0.005\,\mathrm{m/s}$). For larger v the instability is somewhat delayed and oscillations are about the planar solution only (Fig. 6(c,d) for $v = 0.01\,\mathrm{m/s}$), although still of the same amplitude. This amplitude goes down as the speed v is increased further, as illustrated in Fig. 7, where the lateral displacement is tiny. We see again, as in the case of the shallow arch, but now for a completely different failure mode, that speed of the travelling load has a stabilising effect. A few shapes for the swaying arch are given in Fig. 8. The numerical parameters used in these deep-arch runs are: $\Delta s = 0.001$ for $v = 0.005$, $\Delta s = 0.00125$ for $v = 0.01$, $\Delta s = 0.002$ for $v = 0.05$, $\Delta s = 0.00125$ for $v = 0.1$, and $\Delta t = \Delta s/v$ in all cases.

5. Conclusion

We studied the interaction between mass and beam or rod while the former travels along the latter. We first considered a stiff beam undergoing only small deflections. The motion of the mass is then unaffected by the beam vibrations. We showed that the mass triggers small vibrations in the beam. For the case of a more flexible rod undergoing large deformations we then developed an efficient numerical scheme based on Cosserat theory using the generalised α-method for both spatial and temporal discretisation. The method is found to have very good convergence and stability properties (with tunable numerical damping). It can handle discontinuities in the force due to a point load well. Application to a few arch problems shows that a moving load has a stabilising effect on the, otherwise statically unstable, structure, both

Journal of Physics: Conference Series **721** (2016) 012016 doi:10.1088/1742-6596/721/1/012016

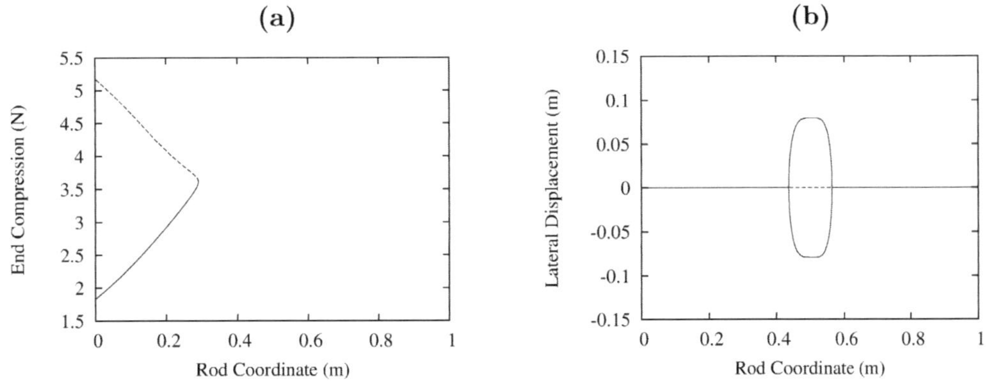

Figure 4. Bifurcation diagrams for the shallow (a) and deep (b) arch under quasistatic load F. Panel (a), showing the end compression $|\boldsymbol{n}(L) \cdot \boldsymbol{k}|$ against the rod coordinate σ for $F = 2.0\,\mathrm{N}$, reveals a fold at $\sigma = 0.291639\,\mathrm{m}$, while panel (b), showing the lateral end displacement x against σ for $F = 3.8\,\mathrm{N}$, reveals a pitchfork bifurcation at $\sigma = 0.436347\,\mathrm{m}$ followed by a reverse pitchfork bifurcation at $\sigma = 0.563653\,\mathrm{m}$. Solid branches are stable, dashed branches are unstable.

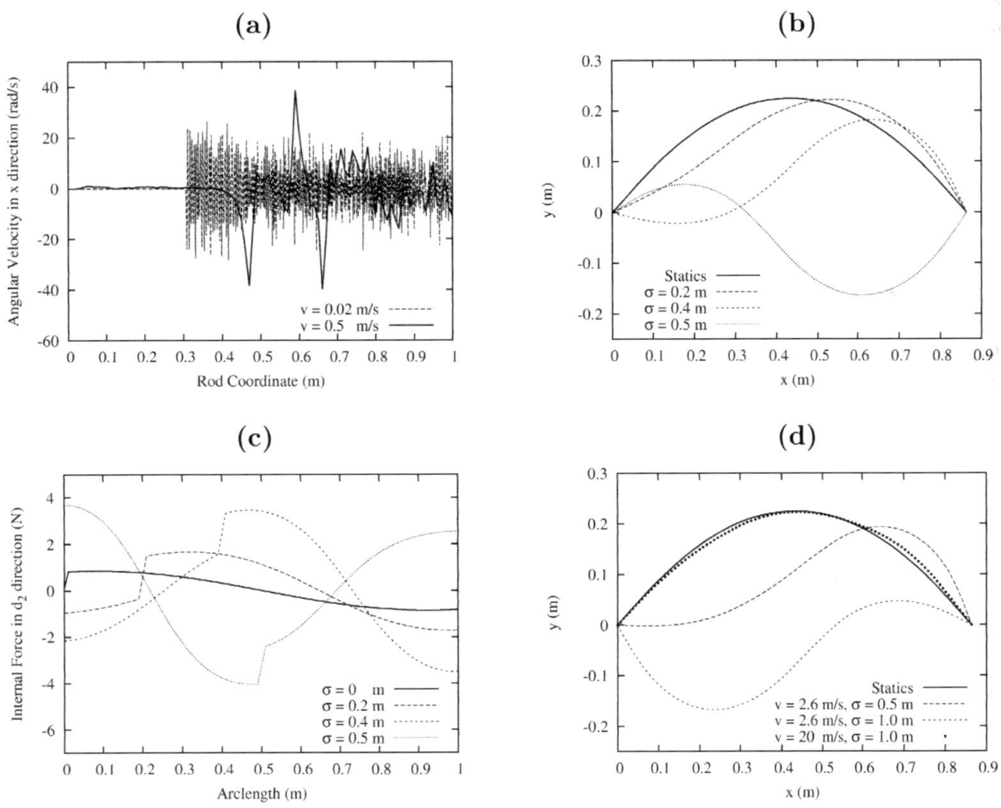

Figure 5. (a) Angular velocity $\boldsymbol{\omega}(L) \cdot \boldsymbol{i}$ about the hinge at $s = L$ as a function of the rod coordinate σ showing delayed buckling as the speed v is increased. (b) Arch shapes at various rod coordinates σ for $v = 0.5\,\mathrm{m/s}$, the last one, at $\sigma = 0.5$, after collapse has occurred. (c) Force component $n_2 = \boldsymbol{n} \cdot \boldsymbol{d}_2$ as a function of arclength s showing the travelling step discontinuity due to the point load F. (d) Arch shapes at higher speeds with collapse suppressed. ($F = 2.0\,\mathrm{N}$.)

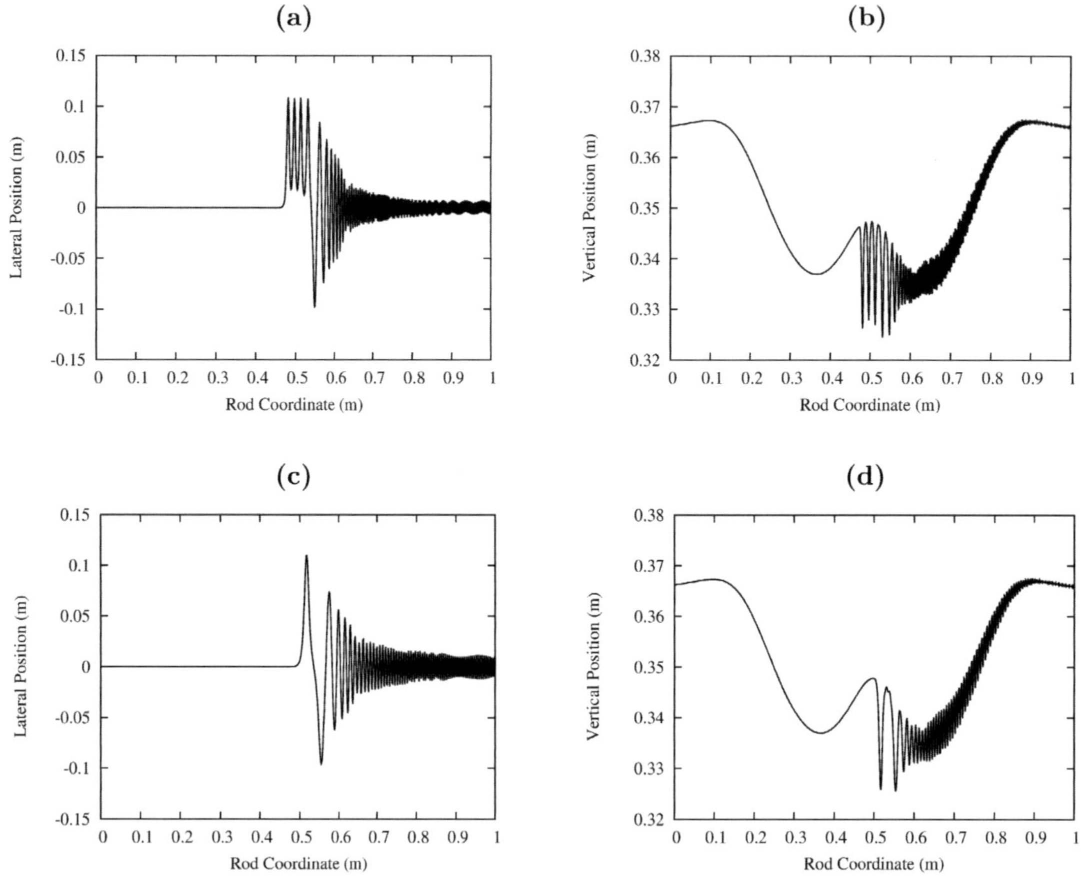

Figure 6. Lateral (x) and vertical (y) position of the deep arch with moving load at speeds $v = 0.005\,\mathrm{m/s}$ (a,b) and $v = 0.01\,\mathrm{m/s}$ (c,d). ($F = 3.8\,\mathrm{N}$.)

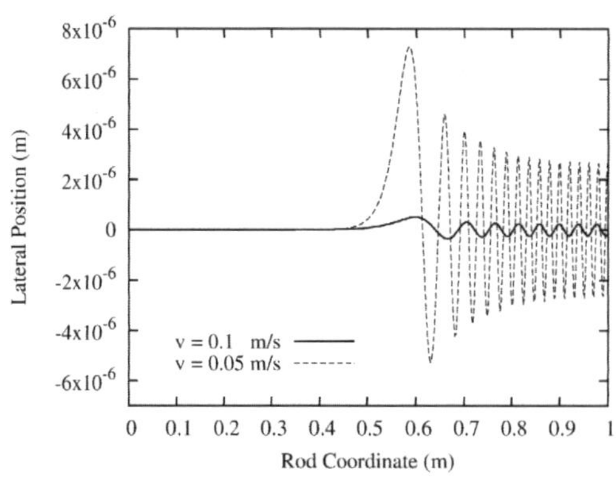

Figure 7. Lateral position x of the deep arch under a load travelling at higher speeds v, showing the stabilising effect of higher speeds. ($F = 3.8\,\mathrm{N}$.)

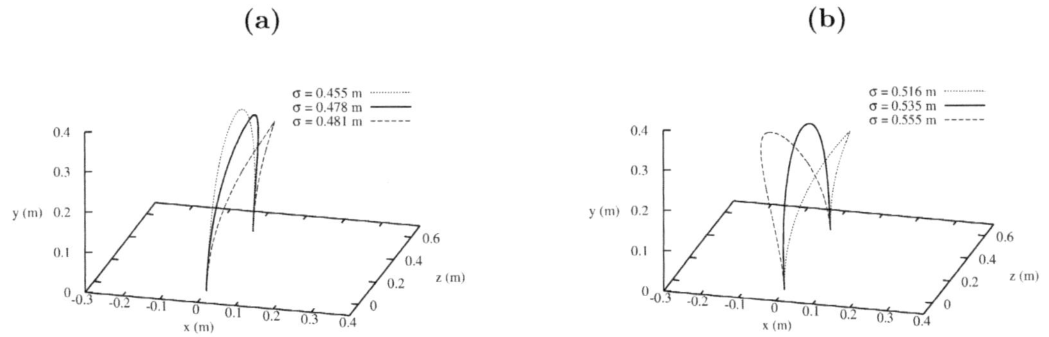

Figure 8. 3D shapes of the deep arch with moving load at different speeds: (a) $v = 0.005\,\text{m/s}$, (b) $v = 0.01\,\text{m/s}$. Shapes are shown at three different rod coordinates. ($F = 3.8\,\text{N}$.)

for in-plane (collapse) instability, for a shallow arch, and for out-of-plane (flop) instability, for a deep arch. In many large-deformation applications of structures carrying a moving mass the mass will move autonomously along the cable, with a varying speed that depends on the local shape and velocity of the cable or arch. In future work we will investigate this two-way coupling between the dynamics of the cable and the mass.

References

[1] Wang P K C and Wei J 1987 Vibrations in a moving flexible robot arm, *J. Sound Vibr.* **116** 149–160
[2] Dwivedy S K and Eberhard K 2006 Dynamic analysis of flexible manipulators, a literature review, *Mech. Mach. Theory* **41** 749–777
[3] Ouyang H 2011 Moving-load dynamic problems: A tutorial (with a brief overview), *Mech. Syst. Signal Proc.* **25** 2039–2060
[4] Fryba L 1972 *Vibration of Solids and Structures under Moving Loads* (Groningen: Noordhoff)
[5] Sadiku S and Leipholz H 1987 On the dynamics of elastic systems with moving concentrated masses, *Arch. Appl. Mech.* **57** 223–242
[6] Ting E C, Genin J and Ginsberg J H 1974 A general algorithm for moving mass problems, *J. Sound Vibr.* **33** (1974) 49–58
[7] Ryu B J, Lee J W, Yim K B and Yoon Y S 2005 Dynamic behaviors of an elastically restrained beam carrying a moving mass, J. Mech. Sci. Technol. **20** 1382–1389
[8] Siddiqui S A Q, Golnaraghi M F and Heppler G R 2003 Large free vibrations of a beam carrying a moving mass, *Int. J. Non-Lin. Mech.* **38** 1481–1493
[9] Wu J J and Whittaker A R 1999 The natural frequencies and mode shapes of a uniform cantilever beam with multiple two-dof spring-mass systems, *J. Sound Vib.* **227** 361–381
[10] Zhao X W, Hu Z D and van der Heijden G H M 2015 Dynamic analysis of a tapered cantilever beam under a travelling mass, *Meccanica* **50** 1419–1429
[11] Antman S S 2005 *Nonlinear Problems of Elasticity* (2nd ed.) (New York: Springer)
[12] Goyal S, Perkins N C and Lee C L 2008 Non-linear dynamic intertwining of rods with self-contact, *Int. J. Non-Lin. Mech.* **43** 65–73
[13] Sobottka G, Lay T and Weber A 2008 Stable integration of the dynamic Cosserat equations with application to hair modeling, *J. WSCG* **16** 73–80
[14] Chung J and Hulbert G M 1993 A time integration algorithm for structural dynamics with improved numerical dissipation – the generalized-α method, *J. Appl. Mech.* **60** 371–375
[15] Nowak U and Weimann L 1991 A family of Newton codes for systems of highly nonlinear equations, *Technical Report TR-91-10* (Konrad-Zuse-Zentrum für Informationstechnik Berlin)
[16] Murray R M, Li Z X and Sastry S S 1994 *A Mathematical Introduction to Robotic Manipulation* (Boca Raton: CRC Press)
[17] Xu F Y, Wang L, Wang X S and Jiang G P 2013 Dynamic performance of a cable with an inspection robot: analysis, simulation, and experiments, *J. Mech. Sci. Technol.* **27** 1479–1492
[18] Olsson M 1991 On the fundamental moving load problem, *J. Sound Vibr.* **145** 299–307